Mathematical Engineering

For further volumes:
http://www.springer.com/series/8445

Ülo Lepik · Helle Hein

Haar Wavelets

With Applications

Springer

Ülo Lepik
Institute of Mathematics
University of Tartu
Tartu
Estonia

Helle Hein
Institute of Computer Science
University of Tartu
Tartu
Estonia

ISSN 2192-4732 ISSN 2192-4740 (electronic)
ISBN 978-3-319-04294-7 ISBN 978-3-319-04295-4 (eBook)
DOI 10.1007/978-3-319-04295-4
Springer Cham Heidelberg New York Dordrecht London

Library of Congress Control Number: 2013957382

Printed on acid-free paper

Springer is part of Springer Science+Business Media (www.springer.com)

Preface

In many disciplines, problems appear which can be formulated with the aid of differential or integral equations. In simpler cases, such equations can be solved analytically, but for more complicated cases, numerical procedures are needed. At present we have, for this purpose, several methods and programs. In recent times, the wavelet-based methods have gained great popularity, where different wavelet families such as Daubechies Coiflet, Symlet, etc., wavelets are applied. A shortcoming of these wavelets is that they do not have an analytic expression. For this reason, differentiation and integration of these wavelets are very complicated and doubts about the expediency of these wavelets in Calculus arise.

From wavelets which have an analytic expression mathematically the simplest are the Haar wavelets, which consist of pairs of piecewise constant functions. Such functions were introduced by Alfred Haar in 1910 and they have been used for solving problems of Calculus only from 1997.

When compared with other methods of solution, the Haar wavelet approach has some preferences, as mathematical simplicity, possibility to implement standard algorithms, and high accuracy for a small number of grid points. The solutions based on the Haar wavelets are usually simpler and faster than in the case of other methods. For these reasons, the Haar wavelets have obtained a great popularity and the number of papers about Haar Wavelets is rapidly increasing. According to the Science Direct in 12.04.13, there were 3,295 publications about Haar wavelets, among these 1,266 items are on differential and integral equations. To the reader it is difficult to find his way among the great number of these publications; therefore, a text-book about the applications of the Haar wavelets in Calculus is extremely necessary. Unfortunately such a book has been missing up to now.

The aim of the present book is to fulfill this gap, even if partially. At present, time different variants of the Haar method exist. It is not reasonable to handle and analyze all of them in detail; it would make the book less understandable and could confuse the reader. Therefore, we have decided to choose a method of solution, which is sufficiently universal and is applicable to solve all the problems by a unit approach. Other treatments will be referred and discussed in the section related papers, which is added to each chapter.

The book is put together on the basis of 19 papers, which we have published in prereviewed international journals. A unit method of solution is applied for solution of a wide range of problems (different types of differential and integral equations, fractional integral equations, optimal control theory, buckling and vibrations of elastic beams). To demonstrate efficiency and accuracy of the proposed method, a number of examples is solved. Mostly test problems, for which the exact solution or solution obtained by other methods is known, are considered.

The book is meant for researchers in applied mathematics, physics, engineering, and related disciplines, also for teachers of higher schools, graduate and postgraduate students. To make the book accessible for a wider circle of readers, some mathematical finesses are left out.

Financial support from the Estonian Science Foundation under Grant ETF 8830 and SF0180008s12 are gratefully acknowledged.

Ülo Lepik
Helle Hein

Contents

Chapter 1
Preliminaries

1.1 Why we Need the Wavelets?

Consider a function of time $f = f(t)$. The Fourier transform of this function is

$$F(\omega) = \int_{-\infty}^{\infty} f(t)e^{-i\omega t}dt. \tag{1.1}$$

Here the transform variable ω represents angular frequency. If the signal $f(t)$ is put together of harmonic components then the frequency diagram $F = F(\omega)$ consists only of sharp peaks. This approach has been successfully applied for solving many problems (especially for signal processing), but it also has an essential disadvantage. The Fourier method analyses the signal over the whole domain, but does not characterize the motion in time.

Let us illustrate this with an example. Consider two motions

$$(i)\ f(t) = \sin 3t + 0.8 \sin 10t \ for\ t \in [0, 10], \tag{1.2}$$

and

$$(ii)\quad f(t) = \begin{cases} \sin 3t & for\ t \in [0, 5), \\ \sin 10t & for\ t \in (5, 10]. \end{cases} \tag{1.3}$$

These motions and their Fourier diagrams are plotted in Fig. 1.1. Although the motions are quite different, their Fourier diagrams are very similar. The only information we get from the Fourier diagram is that in both motions components with the frequencies $\omega_1 = 3$ and $\omega_2 = 10$ dominate.

The other example is from music. If we analyse a piece of music by the Fourier method we can find out which frequencies (notes) occured, but we do not get any information about the melody. To sum up: the Fourier transform localizes the time series in frequency, but the results are completely delocalized in time.

Ü. Lepik and H. Hein, *Haar Wavelets*, Mathematical Engineering,
DOI: 10.1007/978-3-319-04295-4_1, © Springer International Publishing Switzerland 2014

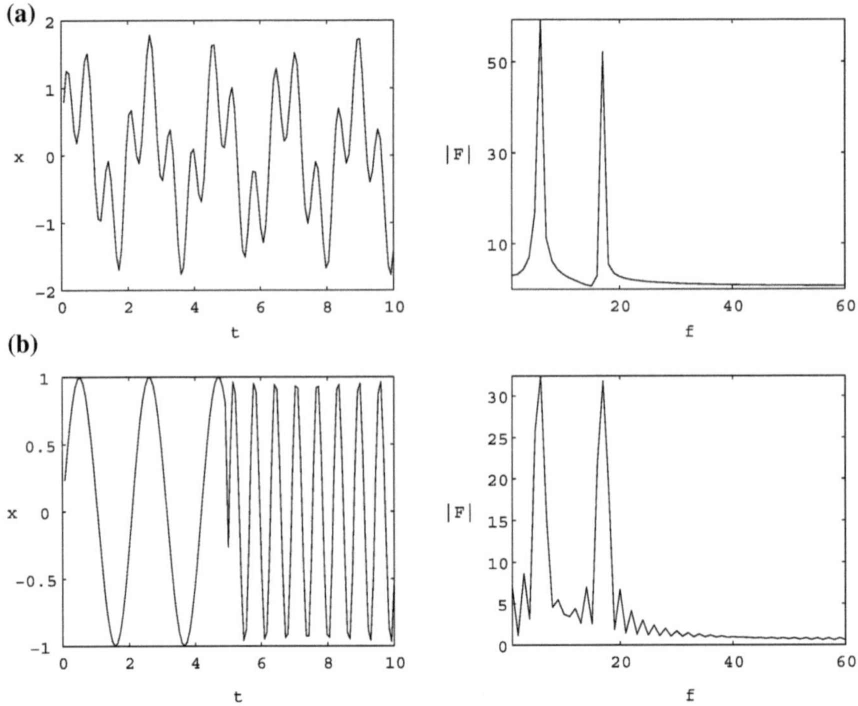

Fig. 1.1 Time history and Fourier diagram. **a** $x = \sin 3t + 0.8 \sin 10t$ for $t \in [0, 10]$ and **b** $x = \sin 3t$ for $t \in [0, 5]$; $x = \sin 10t$ for $t \in [5, 10]$ [7]

Evidently we need a method which analyses our signal over the whole domain and is also able to characterize the motion in time. This can be done with the aid of wavelets.

1.2 Wavelet Families

For generating a wavelet system two basic functions—the scaling function $\varphi = \varphi(t)$ and the mother wavelet $\psi = \psi(t)$ are required. These functions must satisfy the admissibility conditions

$$\int_{-\infty}^{\infty} \frac{\hat{\varphi}(\omega)}{\omega} d\omega < \infty, \quad \int_{-\infty}^{\infty} \frac{\hat{\psi}(\omega)}{\omega} d\omega < \infty. \tag{1.4}$$

Here $\hat{\varphi}(\omega)$ and $\hat{\psi}(\omega)$ denote the Fourier transforms of $\varphi(t)$ and $\psi(t)$, respectively. These conditions imply that if $\hat{\varphi}(\omega)$ and $\hat{\psi}(\omega)$ are smooth then $\hat{\varphi}(0) = \hat{\psi}(0) = 0$.

For getting a wavelet series, which describes the motion both in space and time, we must introduce two parameters. This can be done by assuming

$$\varphi_b(t) = \varphi(t - b),\ \psi_{a,b}(t) = a^{-1/2}\psi(\frac{t - b}{a}). \tag{1.5}$$

The parameter a means rescaling and b-shifting along the t-axis. In some cases it is more convenient to make use of the binary system and take

$$\varphi_k(t) = \varphi(t - k),\ \ \psi_{j,k}(t) = 2^{j/2}\psi(2^j t - k). \tag{1.6}$$

Here j and k are integers; j is called the *dilation parameter*, k—the *translation parameter*.

Next consider a function $f = f(t)$ for which

$$\int_{-\infty}^{\infty} | f(t) |^2 \, dt < \infty. \tag{1.7}$$

This function can be expanded into the wavelet series as follows

$$f(t) = \sum_b c_b\varphi_b(t) + \sum_a \sum_b d_{a,b}\psi_{a,b}(t), \tag{1.8}$$

or in the case of binary system

$$f(t) = \sum_k c_k\varphi_k(t) + \sum_k \sum_j d_{j,k}\psi_{j,k}(t). \tag{1.9}$$

Here the symbols c_b, $d_{a,b}$ (or respectively c_k, $d_{j,k}$) denote the wavelet coefficients. For the case that $\psi_{j,k}(t)$ are orthonormal and $\varphi_k(t)$ is orthogonal to $\psi_{j,k}(t)$ these coefficients can be found according to the formulae

$$\begin{aligned} c_k &= \int f(t)\varphi_k(t)dt, \\ d_{j,k} &= \int f(t)\psi_{j,k}(t)dt, \end{aligned} \tag{1.10}$$

where the integrations are carried out over the whole time domain.

Since we are quite free to choose the scaling function $\varphi(t)$ and mother wavelet $\psi(t)$ different wavelet families can then be generated.

The wavelet transform was introduced in the works of J. Morlet in the early 1980s. From the outset, the wavelet method was considered as a mathematical curiosity but due to extensive research in the 1990s was turned into a well grounded and powerful mathematical tool with many practical applications. Paving the way for this is the paper by Ingrid Daubechies [3].

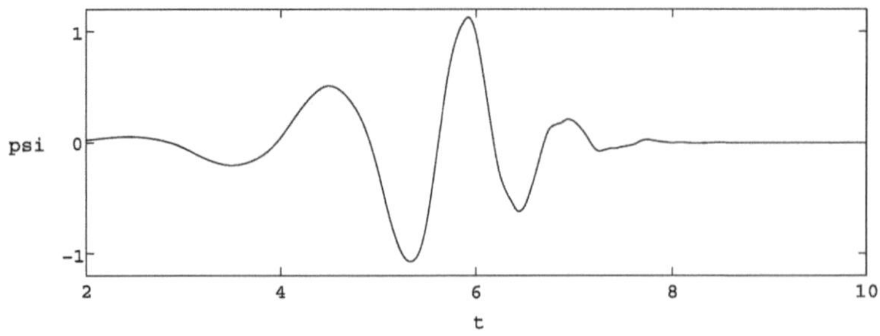

Fig. 1.2 Daubechies mother wavelet of order $J = 6$

The wavelets introduced by Daubechies are quite frequently used for solving different problems. The Daubechies mother wavelet is plotted in Fig. 1.2. These wavelets are differentiable and have a minimum size support.

A shortcoming of the Daubechies wavelets is that they do not have an explicit expression and therefore analytical differentiation or integration is not possible. This complicates the solution of differential equations, where the following type integrals

$$\int_a^b G(t, \psi_{i,k}, \frac{d\psi_{i,k}}{dt}, \frac{d^2\psi_{i,k}}{dt^2}, \ldots)dt \tag{1.11}$$

must be computed (G is generally a nonlinear function). For calculating such integrals the conception of *connection coefficients* is introduced. Calculation of these coefficients is very complicated and must be carried out separately for different types of integrals see e.g. [1, 6]. Besides, it can only be done for some simpler types of nonlinearities (mainly for quadratic nonlinearity). This remark holds also for other types of wavelets (as Symlet, Coiflet, etc. wavelets).

The wavelet method was first applied to solving differential and integral equations in the 1990s. Due to the complexity of the wavelet solutions some pessimistic estimates exist. So Strang and Ngyen in 1996 wrote in their text-book [9] "... the competition with other methods is severe. We do not necessarily predict that wavelets will win" (p. 394). Jameson [4] writes "... nonlinearities etc., when treated in a wavelet subspace, are often unnecessarily complicated ... There appears to be no compelling reason to work with Galerkin-style coefficients in a wavelet method" (p. 1982).

Obviously attempts to simplify solutions based on the wavelet approach are wanted.

Nowadays we possess several wavelet families for which analytic expressions for scaling function and mother wavelet are defined. Since such wavelets have been applied for solving differential and integral equations in several papers, we will describe some of them here.

In the case of the *Morlet wavelets* (also known as Gabor wavelets) the mother wavelet is taken in the form

$$\psi(t) = e^{i\omega_0 t} e^{-0.5t^2}, \; i = \sqrt{-1}, \; \omega_0 \geq 5, \tag{1.12}$$

the scaling function $\varphi(t)$ is not introduced.

The *harmonic wavelet transform* proposed by Newland [8] in 1993 is defined as

$$\varphi(t) = \frac{e^{i2\pi t} - 1}{i2\pi t}, \; \psi(t) = \frac{1}{i2\pi t}(e^{i4\pi t} - e^{i2\pi t}). \tag{1.13}$$

These wavelets are remarkable in the sense that they have a very simple Fourier transform for the scaling function [1].

$$\hat{\varphi}(\omega) = \begin{cases} \frac{1}{2\pi} \text{ for } 0 \leq \omega < 2\pi \;, \\ 0 \text{ elsewhere.} \end{cases} \tag{1.14}$$

The *Shannon wavelets* are defined by [2]

$$\begin{aligned} \varphi(t) &= \mathrm{sinc}\pi t = \tfrac{\sin \pi t}{\pi t}, \\ \psi(t) &= \mathrm{sinc}\tfrac{t}{2} \cos \tfrac{3\pi t}{2}. \end{aligned} \tag{1.15}$$

The benefit of Shannon wavelet is that the functions $\varphi(t)$ and $\psi(t)$ also have a simple Fourier transform

$$\hat{\varphi}(\omega) = \begin{cases} \frac{1}{\sqrt{2\pi}} \text{ for } \mid \omega \mid \leq \pi \;, \\ \quad 0 \text{ otherwise,} \end{cases} \tag{1.16}$$

and

$$\hat{\psi}(\omega) = \begin{cases} \frac{1}{\sqrt{2\pi}} \exp(-i\omega/2) \text{ for } \pi \leq \mid \omega \mid \leq 2\pi \;, \\ \quad 0 \text{ otherwise.} \end{cases} \tag{1.17}$$

In many papers the Legendre wavelets have been applied. For these wavelets different variants exist. We follow here the method proposed for linear Legendre multiwavelets in [5].

For constructing these wavelets two scaling functions are introduced and they are described as

$$\varphi^0(t) = 1, \; \varphi^1(t) = \sqrt{3}(2t - 1), \; 0 \leq t < 1. \tag{1.18}$$

The corresponding mother wavelets are defined as

$$\psi^0(t) = \begin{cases} -\sqrt{3}(4t - 1) \text{ for } 0 \leq t \leq 1/2, \\ \sqrt{3}(4t - 3) \text{ for } 1/2 \leq t < 1, \end{cases} \tag{1.19}$$

and

$$\psi^1(t) = \begin{cases} 6t - 1 \text{ for } 0 \leq t \leq 1/2, \\ 6t - 5 \text{ for } 1/2 \leq t < 1. \end{cases} \tag{1.20}$$

With dilation and translation, we obtain the Legendre wavelets

$$\psi^j_{k,n}(t) = \begin{cases} 2^{k/2}\psi^j(2^k t - n) \text{ for } 2^{-k}n \le t \le 2^{-k}(n+1)\,, \\ 0 \textit{ otherwise}. \end{cases} \tag{1.21}$$

Here $n = 0, 1, \ldots, 2^k - 1$, k is an non-negative integer, $j = 0, 1$. A function $f(t)$ defined over $[0, 1)$ may be expanded as

$$f(t) = c_0\varphi^0(t) + c_1\varphi^1(t) + \sum_{k=0}^{M}\sum_{j=0}^{1}\sum_{n=0}^{2^k-1} c^j_{k,n}\psi^j_{k,n}(t). \tag{1.22}$$

Due to orthonormality of the Legendre wavelets, the wavelet coefficients can be calculated from (1.10). In some papers the Laguerre, Chebyshev or B-spline wavelets are used.

Haar wavelets are based on the functions which were introduced by Hungarian mathematician Alfred Haar in 1910. The Haar wavelets are made up of piecewise constant functions and are mathematically the simplest among all the wavelet families. A good feature of these wavelets is the possibility to integrate them analytically arbitrary times. They can be interpreted as a first order Daubechies wavelet. The Haar wavelets have been applied for solving several problems of mathematical calculus. Some of these results will be discussed and analysed in the following chapters of this book.

References

1. Cattani, C.: Harmonic wavelets towards the solution of nonlinear PDE. Comput. Math. Appl. **50**, 1191–1210 (2005)
2. Cattani, C.: Shannon wavelets theory. Math. Prob. Eng. **121**, 1–24 (2008)
3. Daubechies, I.: Orthonormal bases of compactly supported wavelets. Commun. Pure Appl. Math. **41**, 909–996 (1988)
4. Jameson, L.: A wavelet-optimized, very high order adaptive grid and order numerical method. SIAM: J. Sci. Comput. **19**, 1980–2013 (1998)
5. Khellat, F., Yousefi, S.: The linear Legendre mother wavelets operational matrix of integration and its applications. J. Franklin Inst. **343**, 181–190 (2006)
6. Latto, A., Resnikoff, H., Tenenbaum, E.: The evaluation of connection coefficients of compactly supported wavelets. Proceedings of the French-USA Workshop on Wavelets and Turbulence, pp. 76–87 (1992)
7. Lepik, Ü.: Application of the Haar wavelet transform to solving integral and differential equations. Proc. Est. Acad. Sci. Phys. Math. **56**, 28–46 (2007)
8. Newland, D.: Harmonic wavelet analysis. Proc. R. Soc. **A443**, 203–225 (1993)
9. Strang, G., Nguyen, T.: Wavelets and Filter Banks. Wellesley-Cambridge Press, Cambridge (1996)

Chapter 2
Haar Wavelets

2.1 Haar Wavelets and their Integrals

This section is based on paper [4]. Let us consider the interval $x \in [A, B]$, where A and B are given constants. We define the quantity $M = 2^J$, where J is the maximal level of resolution. The interval $[A, B]$ is divided into $2M$ subintervals of equal length; the length of each subinterval is $\Delta x = (B - A)/(2M)$. Next two parameters are introduced: $j = 0, 1, \ldots, J$ and $k = 0, 1, \ldots, m - 1$ (here the notation $m = 2^j$ is introduced). The wavelet number i is identified as $i = m + k + 1$.

The i-th Haar wavelet is defined as

$$h_i(x) = \begin{cases} 1 & \text{for } x \subset [\xi_1(i), \xi_2(i)), \\ -1 & \text{for } x \in [\xi_2(i), \xi_3(i)), \\ 0 & \text{elsewhere}, \end{cases} \tag{2.1}$$

where

$$\begin{aligned} \xi_1(i) &= A + 2k\mu\Delta x\,, & \xi_2(i) &= A + (2k + 1)\mu\Delta x\,, \\ \xi_3(i) &= A + 2(k + 1)\mu\Delta x\,, & \mu &= M/m\,. \end{aligned} \tag{2.2}$$

These equations are valid if $i > 2$. The case $i = 1$ corresponds to the scaling function: $h_1(x) = 1$ for $x \in [A, B]$ and $h_1(x) = 0$ elsewhere. For $i = 2$ we have $\xi_1(2) = A$, $\xi_2(2) = 0.5(2A + B)$, $\xi_3(2) = B$. The parameters j and k have concrete meaning. The support (the width of the i-th wavelet) is

$$\xi_3(i) - \xi_1(i) = 2\mu\Delta x = (B - A)m^{-1} = (B - A)2^{-j} \tag{2.3}$$

It follows from here that if we increase j then the support decreases (the wavelet becomes more narrow). By this reason it is called the *dilatation parameter*. The other parameter k localises the position of the wavelet in the x-axis; if k changes from 0 to $m - 1$ the initial point of the ith wavelet $\xi_1(i)$ moves from $x = A$ to $x = [A + (m - 1)B]/m$. The integer k is called the *translation parameter*.

Ü. Lepik and H. Hein, *Haar Wavelets*, Mathematical Engineering,

DOI: 10.1007/978-3-319-04295-4_2, © Springer International Publishing Switzerland 2014

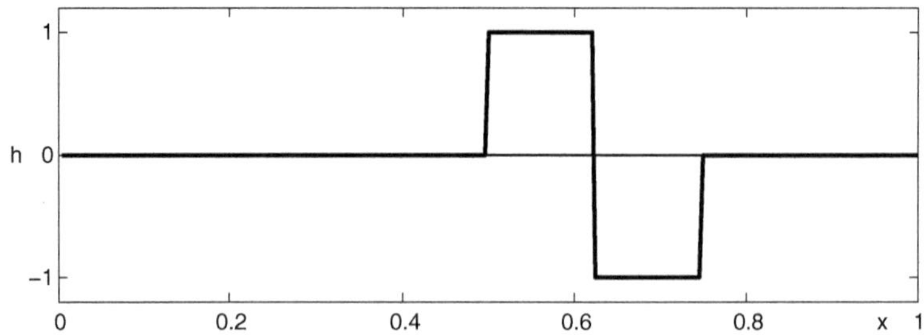

Fig. 2.1 Haar wavelet for $J = 2$; $i = 7$

Let us take an example. If $j = 2$, $J = 2$, $k = 2$, $A = 0$, $B = 1$ we have $m = M = 2^2 = 4$, $\mu = 1$, $\Delta x = 0.125$ and the wavelet number is $i = 7$. According to (2.2) $\xi_1(7) = 0.5, \xi_2(7) = 0.625, \xi_3(7) = 0.75$. This wavelet is plotted in Fig. 2.1.

Eight first wavelets h_1-h_8 are shown in Fig. 2.2.

If the maximal level of resolution J is prescribed then then it follows from (2.1) that

$$\int_A^B h_i(x)h_l(x)dx = \begin{cases} (B-A)2^{-j} & \text{for } l = i, \\ 0 & \text{for } l \neq i. \end{cases} \tag{2.4}$$

So we see that the Haar wavelets are orthogonal to each other.

In the following we need the integrals of the Haar functions

$$p_{\nu,i}(x) = \underbrace{\int_A^x \int_A^x \dots \int_A^x}_{\nu-\text{times}} h_i(t)dt^\nu = \frac{1}{(\nu-1)!}\int_A^x (x-t)^{\nu-1}h_i(t)dt \tag{2.5}$$
$$\nu = 1, 2, \dots, n, \quad i = 1, 2, \dots, 2M.$$

Taking account of (2.1) these integrals can be calculated analytically; by doing it we obtain

$$p_{\alpha,i}(x) = \begin{cases} 0 & \text{for } x < \xi_1(i), \\ \frac{1}{\alpha!}[x-\xi_1(i)]^\alpha & \text{for } x \in [\xi_1(i), \xi_2(i)], \\ \frac{1}{\alpha!}\left\{[x-\xi_1(i)]^\alpha - 2[x-\xi_2(i)]^\alpha\right\} & \text{for } x \in [\xi_2(i), \xi_3(i)], \\ \frac{1}{\alpha!}\left\{[x-\xi_1(i)]^\alpha - 2[x-\xi_2(i)]^\alpha + [x-\xi_3(i)]^\alpha]\right\} & \text{for } x > \xi_3(i). \end{cases} \tag{2.6}$$

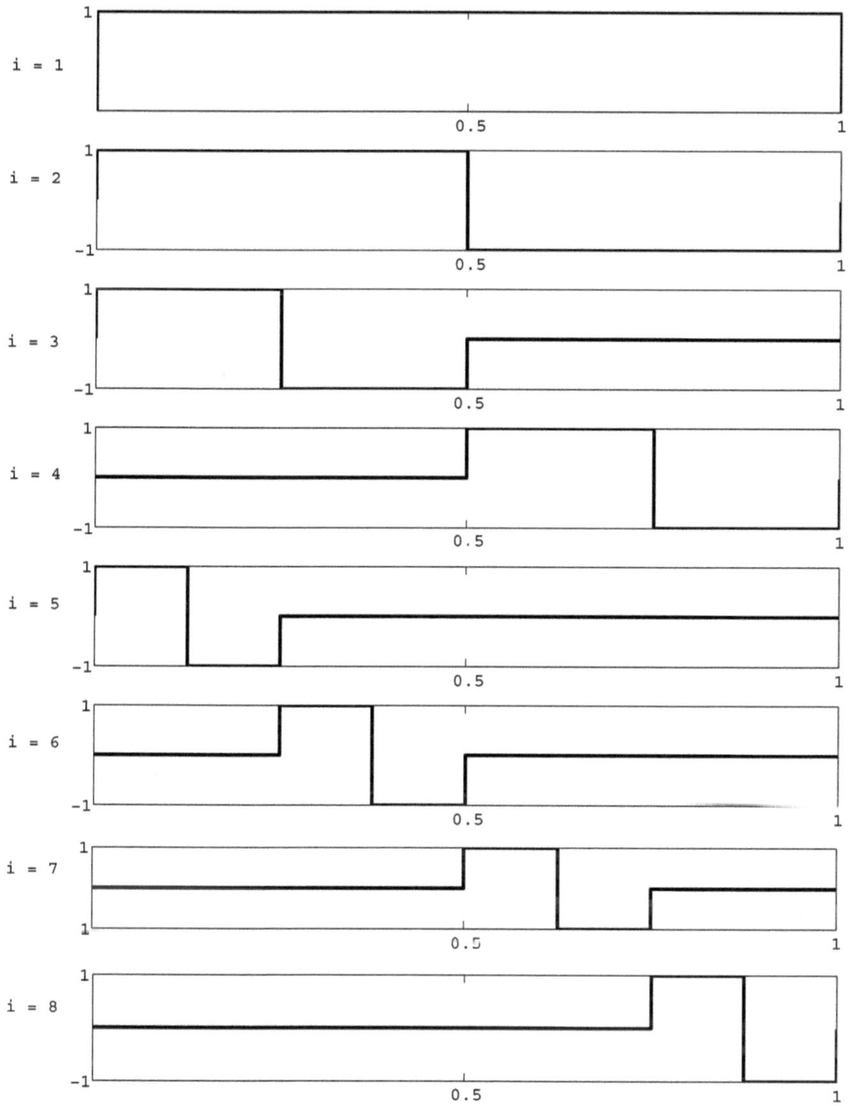

Fig. 2.2 Eight first Haar wavelets

These formulas hold for $i > 1$. In the case $i = 1$ we have $\xi_1 = A, \xi_2 = \xi_3 = B$ and

$$p_{\alpha,1}(x) = \frac{1}{\alpha!}(x - A)^{\alpha}. \tag{2.7}$$

2.2 Haar Matrices

If we want to use the Haar wavelets for the numerical solutions we must put them into a discrete form. There are different ways to do it; in this paper the collocation method is applied.

Let us denote the grid points by

$$\tilde{x}_l = A + l\Delta x, \quad l = 0, 1, \ldots, 2M \tag{2.8}$$

For the collocation points we take

$$x_l = 0.5(\tilde{x}_{l-1} + \tilde{x}_l), \quad l = 1, \ldots, 2M \tag{2.9}$$

and replace $x \to x_l$ in Eqs. (2.1), (2.6) and (2.7). It is convenient to put these results into the matrix form. For this we introduce the Haar matrices $H, P_1, P_2, \ldots, P_\nu$ which are $2M \times 2M$ matrices. The elements of these matrices are $H(i,l) = h_i(x_l)$,$P_\nu(i,l) = p_{\nu i}(x_l)$, $\nu = 1, 2, \ldots$.

For illustration consider the case $A = 0, B = 1, J = 1$. Now $2M = 4$ and the grid points are $\tilde{x}_0 = 0, \tilde{x}_1 = 0.25, \tilde{x}_2 = 0.5, \tilde{x}_3 = 0.75, \tilde{x}_4 = 1$. By calculating the coordinates of the collocation points from (2.9) we find $x_1 = 0.125, x_2 = 0.375, x_3 = 0.625, x_4 = 0.875$. The Haar matrices H, P_1, P_2 are

$$H = \begin{pmatrix} 1 & 1 & 1 & 1 \\ 1 & 1 & -1 & -1 \\ 1 & -1 & 0 & 0 \\ 0 & 0 & 1 & -1 \end{pmatrix}, \quad P_1 = \frac{1}{8}\begin{pmatrix} 1 & 3 & 5 & 5 \\ 1 & 3 & 3 & 1 \\ 1 & 1 & 0 & 0 \\ 0 & 0 & 1 & 1 \end{pmatrix}, \quad P_2 = \frac{1}{128}\begin{pmatrix} 1 & 9 & 25 & 49 \\ 1 & 9 & 23 & 31 \\ 1 & 7 & 8 & 8 \\ 0 & 0 & 1 & 7 \end{pmatrix} \tag{2.10}$$

2.3 Expanding Functions into the Haar Wavelet Series

Consider a square integrable function $f = f(x)$ for $x \in [A, B]$. This function can be expanded into the Haar wavelet series

$$f(x) = \sum_{i=1}^{2M} a_i h_i(x). \tag{2.11}$$

The symbol a_i denotes the Haar wavelet coefficients. The discrete form of (2.11) is (x_l are the collocation points):

$$\hat{f}(x_l) = \sum_{i=1}^{2M} a_i h_i(x_l). \tag{2.12}$$

The matrix form of (2.12) is

$$f = aH. \tag{2.13}$$

Here H is the Haar matrix; a and f are defined as $a = (a_i), f = (f_l)$; both are $2M$ dimensional row vectors. Solving the matrix equation (2.13) with regard to the coefficient vector a we find (H^{-1} denotes the inverse of H)

$$a = fH^{-1}. \tag{2.14}$$

Replacing a into (2.11) we obtain the wavelet approximation of the function $f(x)$ for the level of resolution J. The question arises as to what the degree of exactness of the approximation is (2.11). There are different possibilities to estimate the error function Δ of the wavelet approximations. Here we define the error function as

$$\Delta = \int_A^B \left[f(x) - \hat{f}(x)\right]^2 dx, \tag{2.15}$$

where $\hat{f}(x)$ denotes the approximation of $f(x)$. The discrete form of (2.15) is

$$\Delta_J = \Delta x \sum_{l=1}^{2M} \left[f(x_l) - \hat{f}(x_l)\right]^2. \tag{2.16}$$

The Haar wavelets belong to the group of piecewise constant functions. It is known that if the function is sufficiently smooth, then the convergence rate for the piecewise constant function is $O(M^{-2})$; this result can be transferred also to the Haar wavelet approach. So it could be expected that by doubling the number of collocation points the error roughly decreases four times. Consider two examples.

Example 1: Let $f(x) = \sqrt{x}$ and $x \in (0, 1)$. The Haar matrix is put together as shown in Sect. 2.2. The wavelet coefficients were calculated according to (2.14) and for $J = 3$ they are plotted in Fig. 2.3. Wavelet approximation for some values of J are presented in Fig. 2.4.

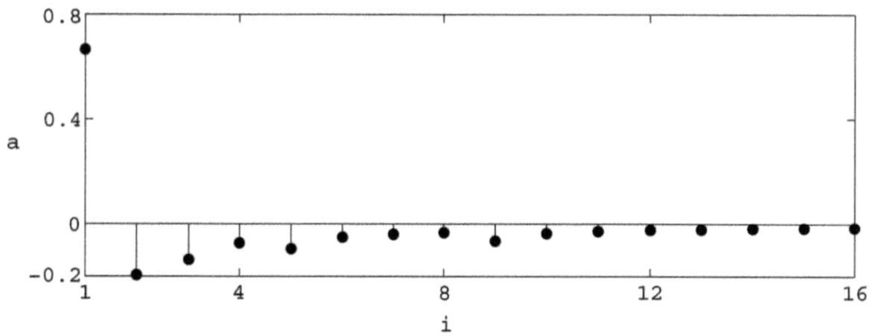

Fig. 2.3 Wavelet coefficients for the equation $f = \sqrt{x},\ x \in (0.1),\ J = 3$

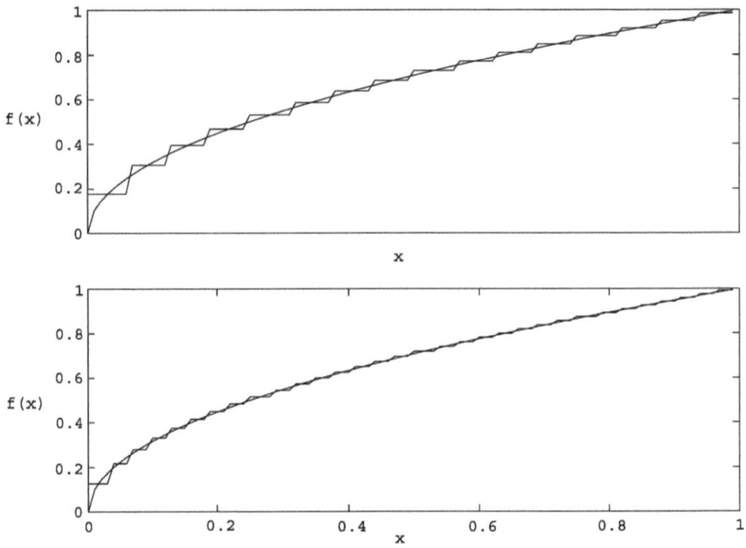

Fig. 2.4 Graph of the function $f = \sqrt{x}$ and its wavelet approximations for (**a**) $J = 3$, (**b**) $J = 4$

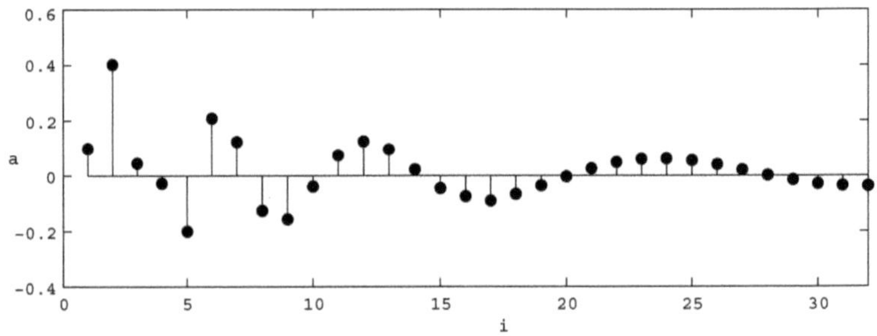

Fig. 2.5 Wavelet coefficients of the function $f = e^{-x} \sin 2\pi x,\ x \in [0, 1]$ for $J = 4$

Example 2: Similar calculations were carried out for $f(x) = e^{-x} \sin 2\pi x, x \in (0, 1)$. The results are plotted in Figs. 2.5 and 2.6.

Error estimates for these two problems are presented in Table 2.1.

It follows from Table 2.1 that the coefficient Δ_{J-1}/Δ_J is near to the predicted theoretical value 4.

From the analysis of these examples, a good feature of the Haar wavelets can be noticed. It follows from Figs. 2.3 and 2.5 that by increasing the wavelet number, the wavelet coefficients rapidly decrease and higher coefficients are practically zero. This obstacle enables to confine to a small number of terms in the wavelet series. The fact that the matrices H and H^{-1} contain many zeros makes the Haar wavelet transform faster when compared with other transforms.

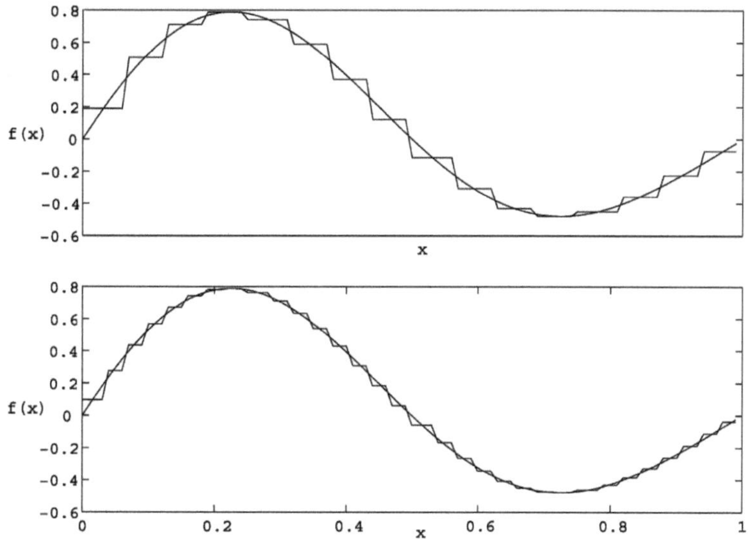

Fig. 2.6 Graph of the function $f = e^{-x}\sin 2\pi x,\ x \in [0, 1]$ and its wavelet approximations for (**a**) $J = 4$, (**b**) $J = 5$

Table 2.1 Error estimates Δ_J for the wavelet approximations

J	$2M$	$f = \sqrt(x)$		$f = exp(-x)\sin x$	
		Δ_J	Δ_{J-1}/Δ_J	Δ_J	Δ_{J-1}/Δ_J
2	8	0.134	–	1.025	3.98
3	16	0.032	3.73	0.259	3.98
4	32	0.010	3.73	0.065	3.98
5	64	0.002	4.25	0.016	4.03

2.4 Non-uniform Haar Wavelets

The present section refers to the paper [5]. Usually it is assumed in the wavelet analysis that the interval $x \in [A, B]$ is distributed into subintervals of equal length. If the function to be expanded into the Haar wavelet series has singularities or the interval (A, B) is infinite, this approach may turn out not to fit (in the class of such problems belong e.g. vibrations under local excitation, boundary value problems, weakly singular integral equations, discontinuities and abrupt changes of the system). In these cases, it is suitable to increase the density of the collocation points in the region of rapid changes. This idea was realized by Dubeau et al. [3] who initiated the theory of non-uniform Haar wavelets. The following analysis is based on the papers of Lepik [4, 5].

We distribute the interval $x \in [A, B]$ optionally into $2M$ subintervals so that $\tilde{x}(0) = A,\ \tilde{x}(2M) = B$ and $\tilde{x}(l + 1) > \tilde{x}(l)$ for $l = 0, 1, \ldots, 2M - 1$.

We define the ith wavelet as

$$h_i(x) = \begin{cases} 1 \text{ for } x \in [\xi_1(i), \xi_2(i)], \\ -c_i \text{ for } x \in [\xi_2(i), \xi_3(i)], \\ 0 \text{ elsewhere.} \end{cases} \tag{2.17}$$

Here the following notations are introduced:

$$\begin{aligned} &\xi_1(i) = x(2k\mu),\ \xi_2(i) = x[(2k+1)\mu], \\ &\xi_3(i) = x[2(k+1)\mu],\ \mu = M/m. \end{aligned} \tag{2.18}$$

The coefficient c_i is calculated from the requirement

$$\int_A^B h_i(x)dx = 0 \tag{2.19}$$

which gives

$$c_i = \frac{\xi_2(i) - \xi_1(i)}{\xi_3(i) - \xi_2(i)}. \tag{2.20}$$

These equations hold if $i > 2$. For the cases $i = 1$ and $i = 2$ we have $\xi_1(1) = A,\ \xi_2(1) = \xi_3(1) = B,\ \xi_1(2) = A,\ \xi_2(2) = \tilde{x}(M),\ \xi_3(2) = B$. By integrating (2.17) α times we obtain

$$p_{\alpha,i}(x) = \begin{cases} 0 & \text{for } x < \xi_i(i), \\ \frac{1}{\alpha!}[x - \xi_1(i)]^\alpha & \text{for } x \in [\xi_1(i), \xi_2(i)], \\ \frac{1}{\alpha!}\left\{[x - \xi_1(i)]^\alpha - (1 + c_i)[x - \xi_2(i)]^\alpha\right\} & \text{for } x \in [\xi_2(i), \xi_3(i)], \\ \frac{1}{\alpha!}\left\{[x - \xi_1(i)]^\alpha - (1 + c_i)[x - \xi_2(i)]^\alpha + c_i[x - \xi_3(i)]^\alpha\right\} & \text{for } x > \xi_3(i). \end{cases} \tag{2.21}$$

The collocation points are defined by (2.9). The Haar function $h(x)$ and the integrals $p_1(x), p_2(x)$ for $J = 2, A = 0, B = 1, x = [0.1\ \ 0.2\ \ 0.3\ \ 0.4\ \ 0.55\ \ 0.7\ \ 0.85]$ are plotted in Fig. 2.7.

2.5 Algorithms and Programs

All computations in this book were carried out with the aid of the MatLab (Matrix Laboratory) programs. The reason for this choice is the fact that many MatLab programs use the matrix representation; this essentially simplifies the programming and saves computing time. For several problems, as in the solution of linear equations, computing eigenvalues, matrix multiplication and inverse matrices etc. special MatLab programs are available. In addition, the graphics of MatLab allows us to prepare Figures of different form.

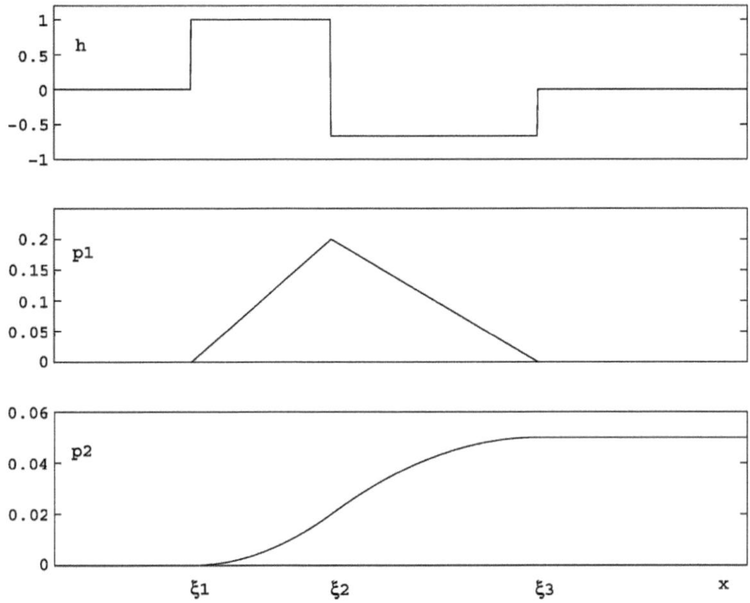

Fig. 2.7 Haar diagrams

There is a difference between the MatLab matrix symbolics and the conventional matrix representation. In MatLab, the vectors are treated as single row matrices, but conventionally as single column matrices. So the MatLab matrix equation $f = aH$, where a and f are row vectors, has in common use the form $\hat{f} = H^T \hat{a}$; here $\hat{a}$, $\hat{f}$ are column vectors and H^T denotes the transverse of H.

In numerical problem solving by the Haar wavelet method, a lot of time is expended for the evaluation of the Haar matrices. Fortunately this process is the same for all problems solved in this book. Therefore it is reasonable to put together universal subprograms for it. Two of these programs are presented in the following.

Program 1: Uniform Haar method: calculation of the Haar matrices.
Code for Program 1.m

```
%Calculation of the integral matrices:uniform case
%Input:resolution level J
%Qutput: Haar  matrices H, P1-P5
M=pow2(J);
M2=2*M;
dX=1/M2;
 for l=1:M2
   X(l)=(l-0.5)*dX;
   H(1,l)=1;
   P1(1,l)=X(l);
```

```
    P2(1,l)=0.5*X(l)^2;
    P3(1,l)=(1/6)*X(l)^3;
    P4(1,l)=(1/24)*X(l)^4;
    P5(1,l)=(1/120)*X(l)^5;
      if (X(l)<0.5)
         H(2,l)=1; P1(2,l)=X(l);
         P2(2,l)=0.5*X(l)^2;
         P3(2,l)=(1/6)*X(l)^3;
         P4(2,l)=(1/24)*X(l)^4;
         P5(2,l)=(1/120)*X(l)^5;
      elseif (X(l)>=0.5)
         H(2,l)=-1;
         P1(2,l)=1-X(l);
         P2(2,l)=0.25-0.5*(1-X(l))^2;
         P3(2,l)=0.25*(X(l)-0.5)+(1/6)*(1-X(l))^3;
         P4(2,l)=(X(l)-0.5)^2/8-(1-X(l))^4/24+1/192;
         P5(2,l)=(X(l)-0.5)^3/24+(1-X(l))^5/120+
         (X(l)-0.5)/192;
       end;end;
  for j=1:J
     m=pow2(j);
  for k1=1:m
       k=k1-1;
       i=m+k1;
       ksi1=k/m;
       ksi2=(k+0.5)/m;
       ksi3=(k+1)/m;
    if ksi3==1
       h(i)=-1; else  end;
   for l=1:M2
    if X(l)<ksi1
       H(i,l)=0; P1(i,l)=0; P2(i,l)=0;P3(i,l)=0;
       P4(i,l)=0; P5(i,l)=0;
     elseif X(l)<ksi2

       H(i,l)=1;
       P1(i,l)=X(l)-ksi1;
       P2(i,l)=0.5*(X(l)-ksi1)^2;
       P3(i,l)=(X(l)-ksi1)^3/6;
       P4(i,l)=(X(l)-ksi1)^4/24;
       P5(i,l)=(X(l)-ksi1)^5/120;
     elseif X(l)<ksi3
       H(i,l)=-1;
       P1(i,l)=ksi3-X(l);
       P2(i,l)=0.25/m^2-0.5*(ksi3-X(l))^2;
```

```
        P3(i,l)=(0.25/m^2)*(X(l)-ksi2)+(ksi3-X(l))^3/6;
        P4(i,l)=(0.125/m^2)*(X(l)-ksi2)^2-
        (ksi3-X(l))^4/24+1/(192*m^4);
        P5(i,l)=(X(l)-ksi2)^3/(24*m^2)+(ksi3-X(l))^
        5/120 +(X(l)-ksi2)/(192*m^4);
      elseif X(l)>=ksi3
        H(i,l)=0;
        P1(i,l)=0;
        P2(i,l)=0.5*((X(l)-ksi1)^2-2*(X(l)-ksi2)^2
        +(X(l)-ksi3)^2);
        P3(i,l)=(0.25/m^2)*(X(l)-ksi2);
        P4(i,l)=(0.125/m^2)*(X(l)-ksi2)^2+1/(192*m^4);
        P5(i,l)=(X(l)-ksi2)^3/(24*m^2)+(X(l)-ksi3)/
        (192*m^4)+1/(384*m^5);
     else;end;end;
   end;end;
```

Program 2: Non-uniform Haar method: calculation of the Haar matrices. Code for Program 2.m

```
%Calculation of the Haar matrices:non-uniform case
%Input: J, A, B; gridpoints x(1), x(2),...,x(M2)
%Output:Haar matrix H, P1 - P4
  co=1/24;
  M=pow2(J);
  M2=2*M;
  ksi1(1)=A; ksi2(1)=B; ksi3(1)=B;
  ksi1(2)=A; ksi2(2)=x(M); ksi3(2)=B;
for j=1:J
   m=pow2(j);
   mu=round(M/m);
   ksi1(m+1)=A;
   ksi2(m+1)=x(mu);
   ksi3(m+1)=x(2*mu);
for k=1:m-1
     i=m+k+1;
     ksi1(i)=x(2*k*mu);
     ksi2(i)=x((2*k+1)*mu);
     ksi3(i)=x(2*(k+1)*mu);
 end;end;
 xc(1)=0.5*(x(1)+A);
for l=2:M2;
   xc(l)=0.5*(x(l)+x(l-1));
   c(l)=(ksi2(l)-ksi1(l))/(ksi3(l)-ksi2(l));
end;
```

```
for i=1:M2
   K(i)=0.5*(ksi2(i)-ksi1(i))*(ksi3(i)-ksi1(i));
for l=1:M2;
   X=xc(l);
if X<ksi1(i);
   H(i,l)=0;
   P1(i,l)=0; P2(i,l)=0;P3(i,l)=0;P4(i,l)=0;
 elseif X<ksi2(i);

   H(i,l)=1;
   P1(i,l)=X-ksi1(i);
   P2(i,l)=0.5*(X-ksi1(i))^2;
   P3(i,l)=(X-ksi1(i))^3/6;
   P4(i,l)=co*(X-ksi1(i))^4;
 elseif X<ksi3(i);
   H(i,l)=-c(i);
   P1(i,l)=c(i)*(ksi3(i)-X);
   P2(i,l)=K(i)-0.5*c(i)*(ksi3(i)-X)^2;
   P3(i,l)=K(i)*(X-ksi2(i))+(ksi3(i)-X)^3/6;
   P4(i,l)=co*((X-ksi1(i))^4-2*(X-ksi2(i))^4);
 elseif X>=ksi3(i);
   H(i,l)=0;
   P1(i,l)=0; P2(i,l)=K(i);
   P3(i,l)=K(i)*(X-ksi2(i));
   P4(i,l)=co*((X-ksi1(i))^4-2*(X-ksi2(i))^4
   +(X-ksi3(i))^4);
else end;
end;
end;
```

2.6 Related Papers

Haar functions belong to a group of rectangular waves with magnitude ± 1 in some intervals and zeros elsewhere. From this group, besides Haar wavelets, the most known are the *block-pulse functions* and *Walsh functions*.

Let us divide the interval $x \in [0, L]$ into N parts of equal length and denote $\Delta x = L/N$. The block-pulse functions (BPF) are defined in the interval $x \in [0, L]$ by

$$b_i(x) = \begin{cases} 1 \text{ for } (i-1)\Delta x \leq x < i\Delta x\,, \\ 0 \; elsewhere, \;\; i = 1, 2, \ldots, N. \end{cases} \tag{2.22}$$

Since the block functions form a complete set of orthogonal functions then an arbitrary function $f(x)$ can be expanded by

$$f(x) = CB(x). \tag{2.23}$$

Here C denotes the coefficients vector $C = [c_1, c_2, \ldots, c_N]^T$ and $B(x) = [b_1(x), b_2(x), \ldots, b_N(x)]^T$.

Taking into account the orthogonality of the BPF we find

$$c_i = \frac{1}{L}\int_0^L f(x)b_i(x)dx. \tag{2.24}$$

The system of Walsh functions $w_n(x)$ is defined on [0, 1] by

$$w_n(x) = (-1)^S, \tag{2.25}$$

where

$$S = \sum_{i=0}^{k} \sigma_{i+1} n_{-i}, \tag{2.26}$$

Here $\sigma_1, \sigma_2, \ldots$ is the dyadic expansion of x with infinitely many zeros and

$$n = n_0 + 2n_{-1} + \cdots + 2^k n_{-k} \tag{2.27}$$

is the dyadic expansion of a positive integer n.

Several authors have made use of the block-pulse and Walsh functions for solving differential and integral equations.

Xiang et al. [9] proposed a Haar-type orthogonal matrix (HTOM), which contains Haar, Walsh and other orthogonal matrices intervenient of the former two. The authors assert that HTOM is very suitable for engineering applications.

Stankovic and Falkowski [8] proposed different generalizations of the Haar transform as sign Haar transform, Haar functions, p-adic groups, Zhang-Moraga functions.

Next the epoch-making paper by Chen and Hsiao [2] is referenced (note that here the symbols may have different meaning as in the present book).

Let us denote the Haar functions by $h_i(t), i = 0, 1, \ldots, m-1$ and define the m-dimensional column vector

$$H_m(t) = [h_1(t), h_2(t), \ldots, h_{m-1}(t)]^T. \tag{2.28}$$

The integral of this vector will be marked as

$$P_m H_m = \int_0^1 H_m(t)dt. \tag{2.29}$$

Here P_m is a $m \times m$ matrix, which is called the *operational matrix of integration*, it can be calculated as

$$P_m = [\int_0^1 H_m(t)dt]H_m^{-1}. \tag{2.30}$$

If $m = 1$ or $m = 2$ we find $P_1 = [0.5]$, $P_2 = 0.25[2\ -1;\ 1\ 0]$. Chen and Hsiao proposed a formula which allows successively evaluate higher order matrices. For this purpose the m-square matrix P_m is divided into four submatrices

$$P_m = \begin{pmatrix} P_{am} & P_{bm} \\ P_{cm} & P_{dm} \end{pmatrix}, \tag{2.31}$$

where

$$\begin{aligned} P_{am} &= P_{m/2}, \quad P_{bm} = -\tfrac{1}{2m}H_{m/2}, \\ P_{cm} &= \tfrac{1}{2m}H_{m/2}^{-1}, \ P_{dm} = 0. \end{aligned} \tag{2.32}$$

If the matrices $P_{m/2}$, $H_{m/2}$ are known we can easily put together the higher order matrix P_m. This approach has been applied in several papers.

Maleknejad and Mirzaee [6] used the term *rationalized Haar wavelets* which should accent the fact that to be compared with the earlier conception about Haar wavelets, the rational numbers were deleted and the integral powers of two were introduced. The difference between the ordinary and rationalized Haar wavelet is mainly symbolic—the conventional notation for Haar wavelets H is replaced by RH.

Aziz et al. [1] and Siraj-ul-Islam et al. [7] investigated integrals of the type

$$\int_A^B f(x)h_i(x)dx \tag{2.33}$$

They worked out a procedure for which numerical integration of such integrals is avoided. This approach is valuable for more complicated functions $f(x)$.

References

1. Aziz, I., Siraj-ul-Islam, Khan, W.: Quadrature rules for numerical integration based on Haar wavelets and hybrid functions. Comput. Math. Appl. **61**, 2770–2781 (2011)
2. Chen, C., Hsiao, C.: Haar wavelet method for solving lumped and distributed-parameter systems. IEEE Proc. Control Theory Appl. **144**, 87–94 (1997)
3. Dubeau, F., Elmejdani, S., Ksantini, R.: Non-uniform Haar wavelets. Appl. Math. Comput. **159**, 675–693 (2004)
4. Lepik, Ü.: Numerical solution of evolution equations by the Haar wavelet method. Appl. Math. Comput. **185**, 695–704 (2007)
5. Lepik, Ü.: Haar wavelet method for solving higher order differential equations. Int. J. Math. Comput. **1**, 84–94 (2008)
6. Maleknejad, K., Mirzaee, B.: Using rationalized Haar wavelet for solving linear integral equations. Appl. Math. Comput. **160**, 579–589 (2005)
7. Siraj-ul-Islam, Aziz, I., Hag, F.: A comparative study of numerical integration based on Haar wavelets and hybrid functions. Comput. Math. Appl. **59**, 2026–2038 (2010)
8. Stankovic, R., Falkowski, B.: The Haar wavelet transform: its status and achievements. Comput. Electr. Eng. **29**, 25–44 (2003)
9. Xiang, X., Zhou, J., Yang, J., Liu, L., An, X., Li, C.: Mechanical signal analysis based on the haar-type orthogonal matrix. Expert Syst. Appl. **36**, 9674–9677 (2009)

Chapter 3
Solution of Ordinary Differential Equations (ODEs)

3.1 Initial Value Problem for n-th Order ODE

Chen and Hsiao [6] considering a first order ODE recommended to expand into the Haar wavelet series not the function to be calculated but its derivative. This suggestion has been very fruitful and nowadays is applied in most cases for the solution of ODEs by the Haar wavelet method.

This idea is applicable also for the solution of higher order differential equations. Here we expand into the Haar wavelet series the highest derivative appearing in the differential equation. The lower order derivatives and the function itself are obtained through integration and the ingredients are then incorporated into the differential equation. Consider the n-th order linear differential equation

$$\sum_{\nu=0}^{n} A_\nu(x) y^{(\nu)}(x) = f(x), \quad x \in [\alpha, \beta] \tag{3.1}$$

with the initial conditions

$$y^{(\nu)}(A) = y_0^{(\nu)}, \quad \nu = 0, 1, \dots, n-1. \tag{3.2}$$

Here $A_\nu(x)$, $f(x)$ are prescribed functions, $y_0^{(\nu)}$ given constants.
The Haar wavelet solution is sought in the form

$$y^{(n)}(x) = \sum_{i=1}^{2M} a_i h_i(x), \tag{3.3}$$

where a_i are the wavelet coefficients.

By integrating (3.3) $n - \nu$ times we obtain

Ü. Lepik and H. Hein, *Haar Wavelets*, Mathematical Engineering,
DOI: 10.1007/978-3-319-04295-4_3,

$$y^{(\nu)}(x) = \sum_{i=1}^{2M} a_i p_{n-\nu,i}(x) + Z_\nu(x)\,, \tag{3.4}$$

where

$$Z_\nu(x) = \sum_{\sigma=0}^{n-\nu-1} \frac{1}{\sigma!}(x-A)^\sigma y_0^{(\nu+\sigma)}\,. \tag{3.5}$$

We shall satisfy (3.1), (3.3)–(3.5) in the collocation points $x(l)$. If we substitute $x \to x(l)$ and replace (3.3)–(3.5) into Eq. (3.1) we get a system of linear equations for calculating the wavelet coefficients a_i. After solving this system, the wanted function $y = y(x)$ is calculated from (3.4) by assuming $\nu = 0$.

It is essential to estimate the exactness of the achieved results. For this purpose we introduce the following error estimates.

(i) If the exact solution $y = y_{ex}(x)$ is known, we define the error estimates as

$$\delta_{ex} = \max_l \left| \frac{y(x_l)}{y_{ex}(x_l)} - 1 \right| \quad \text{(local estimate)} \tag{3.6}$$

and

$$\sigma_{ex} = \parallel y - y_{ex} \parallel /(2M) \quad \text{(global estimate)}\,. \tag{3.7}$$

(ii) If the exact solution is unknown, we define the error vector

$$\varepsilon(l) = \sum_{\nu=0}^{n} A_\nu(x_l) y^{(\nu)}(x_l) - f(x_l)\,. \tag{3.8}$$

The smaller the $\parallel \varepsilon(l) \parallel /(2M)$ the more exact the solution is ($\parallel \varepsilon \parallel = 0$ in the case of the exact solution).

If $\parallel \varepsilon(l) \parallel /(2M)$ is not small enough, we have to go to a higher level of resolution J and repeat the calculations.

Example 3.1: Solve the equation

$$y^{(6)} + y = \sin\frac{3}{2}x \sin\frac{1}{2}x, \quad x \in [0, \pi] \tag{3.9}$$

for the initial conditions $y_0 = 1$, $y_0' = -1$, $y_0'' = 0.3$, $y_0''' = 0$, $y_0^{IV} = 0$, $y_0^{V} = 1$.

According to (3.3)–(3.5) we find

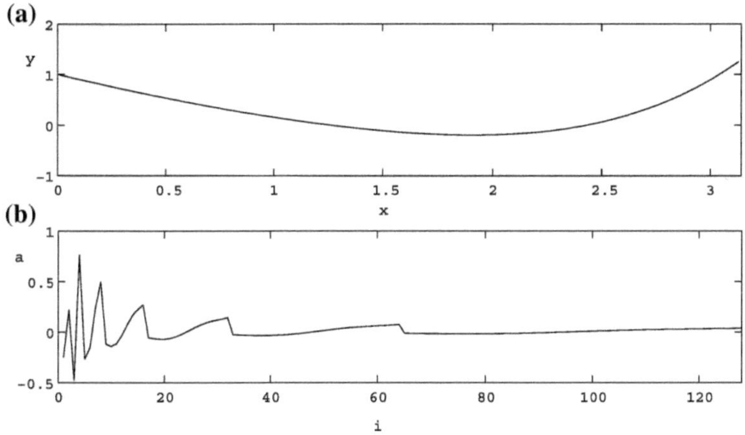

Fig. 3.1 Solution of (3.9) for $J = 6$. **a** Function $y = y(x)$; **b** wavelet coefficients [19]

$$y^{(6)}(x_l) = \sum_{i=1}^{2M} a_i h_i(x_l)$$

$$y(x_l) = \sum_{i=1}^{2M} a_i p_{6,i}(x_l) + \sum_{\sigma=0}^{6} \frac{1}{\sigma!} y_0^{(\sigma)} x_l^{\sigma} . \tag{3.10}$$

Replacing these results into (3.9) we get

$$\sum_{i=1}^{2M} a_i [h_i(x_l) + p_{6,i}(x_l)] + Z(l) = F(l) , \tag{3.11}$$

where $F(l) = \sin \frac{3}{2} x_l \sin \frac{1}{2} x_l$.

The matrix form of (3.11) is

$$a(H + P_6) = F - Z. \tag{3.12}$$

For the "exact solution", the Runge-Kutta solution, realized by the MatLab program ODE45 was taken. Computer simulation gave the following error estimates

$$\begin{aligned}
&(i)\ \delta_{ex} = 3.3E-3,\ \sigma_{ex} = 4.1E-4,\ \text{for } J = 3\\
&(ii)\ \delta_{ex} = 8.3E-4,\ \sigma_{ex} = 5.2E-5,\ \text{for } J = 4\\
&(iii)\ \delta_{ex} = 2.7E-4,\ \sigma_{ex} = 6.5E-6,\ \text{for } J = 5.
\end{aligned}$$

The wavelet solution $y = y(x)$ is plotted in Fig. 3.1a; already in the case $J = 3$ (16 collocation points) it visually coincides with the Runge-Kutta solution. High accuracy of the wavelet results for a small number of calculation points is caused by the fact that by increasing the wavelet number i, the coefficients a_i rapidly decrease (Fig. 3.1b).

3.2 Treatment of Boundary Value Problems

Let us consider once more (3.1). We assume that $\mu < m$ are given the initial conditions (specified at $x = A$), but the remaining $n - \mu$ conditions are prescribed in some internal points $x_1 < B$ or at the boundary $x = B$.

Consider the boundary condition $y^{(\nu)}(x_*) = y_*^{(\nu)}$, where $y_*^{(\nu)}$ is a given number. It follows from (3.4)–(3.5)

$$\begin{aligned} &\sum_{i=1}^{2M} a_i p_{n-\nu,i}(x_*) + Z_\nu(x_*) = y_*^{(\nu)} \\ &Z_\nu(x_*) = \sum_{\sigma=0}^{n-\nu-1} \frac{1}{\sigma!}(x_* - A)^\sigma y_0^{(\nu+\sigma)} . \end{aligned} \tag{3.13}$$

The functions $p_{n-\nu,i}(x_*)$ are calculated from (2.6) replacing there $x \to x_*$. Such an analysis is carried out for each boundary condition. By doing this we get a system of $n-\nu$ linear equations from which the missing $n-\nu$ initial values can be calculated (of course these formulas also contain the wavelet coefficients a_i).

The following course of solution is the same as demonstrated in Sect. 3.1. Details are explained with the help of the following Example.

Example 3.2: Solve the equation

$$xy^{IV} + 5y''' = 24, \quad x \in [1, 3] \tag{3.14}$$

for $y(1) = 0$, $y''(1) = 200$, $y(3) = 0$, $y'''(3) = 0$.

Making use of (3.3)–(3.5) we find

$$\begin{aligned} y^{IV}(x) &= \sum_{i=1}^{2M} a_i h_i(x) \\ y'''(x) &= \sum_{i=1}^{2M} a_i p_{1,i}(x) + y'''(1) \\ y(x) &= \sum_{i=1}^{2M} a_i p_{4,i}(x) + \frac{1}{6} y'''(1)(x-1)^3 + \frac{1}{2} y''(1)(x-1)^2 \\ &\quad + y'(1)(x-1) + y(1). \end{aligned} \tag{3.15}$$

Taking in view the initial and boundary conditions we find from (3.15):

$$
\begin{aligned}
y(3) &= \sum_{i=1}^{2M} a_i p_{4,i}(3) + 2y'(1) + 400 + \frac{4}{3} y'''(1) = 0 \\
y'''(3) &= \sum_{i=1}^{2M} a_i p_{1,i}(3) + y'''(1) = 0\,.
\end{aligned}
\tag{3.16}
$$

From this system we calculate the values of the missing initial conditions:

$$
\begin{aligned}
y'''(1) &= -\sum_{i=1}^{2M} a_i p_{1,i}(3) \\
y'(1) &= \sum_{i=1}^{2M} a_i \left[\frac{2}{3} p_{1,i}(3) - \frac{1}{2} p_{4,i}(3) \right] - 200\,.
\end{aligned}
\tag{3.17}
$$

It follows from (3.4) that

$$
y'''(x) = \sum_{i=1}^{2M} a_i p_{1,i}(x) + y'''(1)\,. \tag{3.18}
$$

Substituting all these results into (3.1) and discretizising the result $x \to x_l$ we get a system of linear equations for calculating the wavelet coefficients a_i:

$$
\sum_{i=1}^{2M} a_i \left[h_i(x_l) x_l + 5 p_{1,i}(x_l) - 5 p_{1,i}(3) \right] = 24, \quad l = 1, 2, \ldots, 2M. \tag{3.19}
$$

After solving (3.19) we evaluate the function $y = y(x)$ from (3.15).
Exact solution of (3.9) is

$$
y_{ex}(x) = \frac{1}{5} \left(243/x^2 - 1026 + 1020x - 241x^2 + 4x^3 \right)\,. \tag{3.20}
$$

Computer simulation was carried out for some values of J. The error estimates were

$$
\begin{aligned}
&\delta_4 = 0.088, \quad \sigma_4 = 2.7E-3 \\
&\delta_5 = 0.021, \quad \sigma_5 = 3.3E-4 \\
&\delta_6 = 0.0054, \quad \sigma_6 = 4.2E-5.
\end{aligned}
$$

The solution for $J = 5$ is plotted in Fig. 3.2a. In the case $J = 6$ our solution and the exact solution visually coincide.

The wavelet coefficients for $J = 5$ are plotted in Fig. 3.2b. It follows from this Figure that the values a_i are significantly bigger than for Example 3.1; this is the reason why the rate of convergence here is somewhat smaller than for Example 3.1.

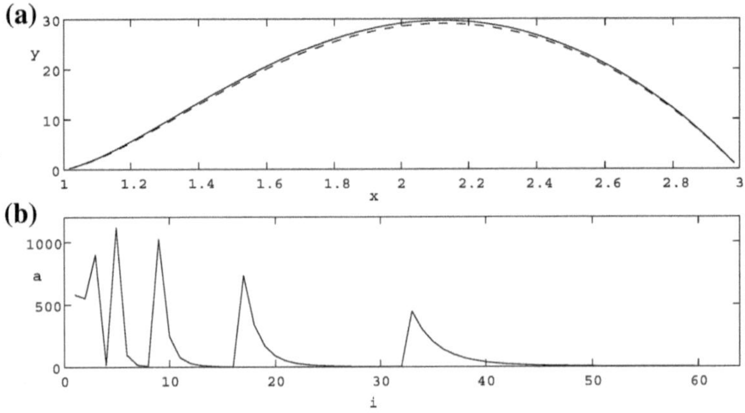

Fig. 3.2 Solution of (3.14) for $J = 5$. **a** Haar wavelet solution—exact solution; **b** wavelet coefficients [19]

3.3 A Modified Solution

If we expand into the Haar series the highest derivative $y^{(n)}$ then the Haar wavelet approximation for $y^{(n)}$ would not be continuous (it consists of piecewise constant segments). In some cases, continuity of $y^{(n)}$ could be needed. As an example consider the equation of motion of a particle

$$my'' = f(x, y, y'). \tag{3.21}$$

Here x denotes time, m—mass, my'' is the acting force. By some physical reasons continuity of the force may be needed.

If we want to guarantee the continuity of the highest derivative $y^{(n)}$ then we should expand into Haar series the derivative $y^{(n+\nu)}$, where $\nu = 1, 2, \ldots$. This idea was realized in [20] . Most important is the case $\nu = 1$ and we confine us in the following to it. The modified solution is smoother than the conventional Haar function; therefore here some decrease of the error function may be expected. Let us clarify the details of the solution by this method with the aid of two examples.

Example 3.3: Consider the initial value problem

$$y^{(6)} + y = \sin\frac{3}{2}x \sin\frac{x}{2}, \quad x \in [0, \pi] \tag{3.22}$$

for the initial conditions $y_0 = 1$, $y_0' = -1$, $y_0'' = 0.3$, $y_0''' = 0$, $y_0^{IV} = 1$, $y_0^V = 1$. Calculating from (3.22) the complementary initial condition $y^{(6)}(0)$ we find $y^{(6)}(0) = -1$. Now we shall take

Table 3.1 Error estimates for the problem (3.22)

$2M$	δ		σ	
	$\mu = 0$	$\mu = 1$	$\mu = 0$	$\mu = 1$
8	3.3E-3	2.5E-4	4.1E-4	3.2E-5
16	8.3E-4	6.4E-5	5.2E-5	4.0E-6
32	2.7E-4	1.6E-5	6.5E-6	5.1E-7

$$y^{(7)}(x) = \sum_{i=1}^{2M} a_i h_i(x)\,.$$

By integrating it seven times and replacing the results into (3.22) we obtain

$$\sum_{i=1}^{2M} a_i[p_{1,i}(x_l) + p_{7,i}(x_l)] = 1 + Z(l) + F(l), \tag{3.23}$$

where

$$Z(l) = \sum_{\sigma=0}^{6} \frac{1}{\sigma!} y_0^{(\sigma)} x_l^{\sigma}\,, \qquad F(l) = \sin\frac{3}{2}x_l \sin\frac{x_l}{2}\,. \tag{3.24}$$

From (3.23) the wavelet coefficients are calculated. The function to be sought is

$$y(x_l) = \sum_{i=1}^{2M} a_i\, p_{7,i}(x_l) + Z(l)\,. \tag{3.25}$$

Computer simulation for this case was carried out and the results were compared with the conventional Haar wavelet method for which $\mu = 0$. Error estimates (3.6) and (3.7) for both cases are presented in Table 3.1.

It follows from Table 3.1 that the error estimates for $\mu = 1$ are about 10 times lesser than for $\mu = 0$.

Example 3.4: Solve the boundary value problem for

$$x^2 y''' + 6xy'' + 6y' = 6, \quad x \in [2, 4] \tag{3.26}$$

for $y(2) = y(4) = 0$, $y'(3) = 0$.

First consider the conventional case $\mu = 0$, where the solution is sought in the form (3.3). By integrating this equation we find

$$\begin{aligned} y'(x) &= \sum_{i=1}^{2M} a_i p_{2,i}(x) + (x-2)y_0'' + y_0' \\ y(x) &= \sum_{i=1}^{2M} a_i p_{3,i}(x) + \frac{1}{2}(x-2)^2 y_0'' + (x-2)y_0' + y_0 . \end{aligned} \tag{3.27}$$

Satisfying the boundary conditions $y(4) = y'(3) = 0$ we get the system

$$\begin{aligned} 0 &= \sum_{i=1}^{2M} a_i P_2(i,3) + y_0'' + y_0' \\ 0 &= \sum_{i=1}^{2M} a_i P_3(i,4) + 2y_0'' + 2y_0' . \end{aligned} \tag{3.28}$$

Here the notation

$$P_\alpha(i, x_*) = p_{\alpha,i}(x = x_*) \tag{3.29}$$

is introduced.

From the system (3.28) we should evaluate the missing initial values y_0', y_0'', but in the present case it is not possible. From here the conclusion, that for $\mu = 0$ the Haar wavelet method is unable to solve the set up boundary value problem, can be made.

Let us pass to the case $\mu = 1$. Now we assume

$$y^{IV} = \sum_{i=1}^{2M} a_i h_i(x) . \tag{3.30}$$

Making use of (3.4) and (3.5) we find

$$y^{(4-\nu)}(x) = \sum_{i=1}^{2M} a_i p_{\nu,i}(x) + \sum_{\sigma=0}^{\nu-1} \frac{1}{\sigma!} y_0^{(4+\sigma-\nu)} (x-2)^\sigma , \quad \nu = 1, 2, 3, 4. \tag{3.31}$$

Instead of (3.28) we get now the system

$$\begin{aligned} 0 &= \sum_{i=1}^{2M} a_i P_3(i,3) + \frac{1}{2} y_0''' + y_0'' + y_0' \\ 0 &= \sum_{i=1}^{2M} a_i P_4(i,4) + \frac{4}{3} y_0''' + 2y_0'' + 2y_0' . \end{aligned} \tag{3.32}$$

Satisfying (3.26) at the boundary $x = 2$ we obtain

Table 3.2 Error estimates for the problem (3.26)

$2M$	δ	σ
8	1.3E-3	4.7E-5
16	1.7E-4	4.2E-6
32	7.8E-5	1.2E-6
64	2.2E-5	2.3E-7

$$4y_0''' + 12y_0'' + 6y_0' = 6. \tag{3.33}$$

From (3.32) and (3.33) the missing initial values are calculated; by doing this we find

$$\begin{aligned} y_0' &= \sum_{i=1}^{2M} a_i \left[P_4(i,4) - 4P_3(i,3)\right] - 2 \\ y_0'' &= \frac{1}{2}\sum_{i=1}^{2M} a_i P_4(i,4) + 2 \\ y_0''' &= -3\sum_{i=1}^{2M} a_i \left[P_4(i,4) - 2P_3(i,3)\right]. \end{aligned} \tag{3.34}$$

Next the results (3.32), (3.35) are substituted into (3.26). Satisfying this equation in the collocation points x_l we get for a_i the system

$$\sum_{i=1}^{2M} a_i S(i,l) = -12(x_l - 2), \quad (l = 1, 2, \ldots, 2M), \tag{3.35}$$

where

$$\begin{aligned} S(i,l) = {}& p_{1,i}(l)x_l^2 + 6p_{2,i}(l)x_l + 6p_{3,i}(l) + 12(5x_l^2 - 12x_l + 4)P_3(i,3) \\ & - 6(5x_l^2 - 13x_l + 6)P_4(i,4). \end{aligned} \tag{3.36}$$

The wanted function $y(x)$ is calculated from (3.31) by taking $\nu = 4$.
This boundary value problem has an exact solution

$$y_{ex} = x - \frac{15}{2} + \frac{17}{x} - \frac{12}{x^2}. \tag{3.37}$$

The error estimates (3.26) are presented in Table 3.2.
It follows from Table 3.2 that the exactness of the obtained solutions is rather good: already in the case of $2M = 8$ calculation points, the curves of the exact and wavelet solutions visually coincide.

3.4 Benefits of the Non-uniform Haar Wavelets

There exist some problems for which the conventional Haar wavelet method may be unsuitable. Here belong e.g. differential equations under the local excitation, boundary layer problems and asymptotic behavior of solutions, etc. In such cases it is reasonable to increase the number of the collocation points in subregions of rapid change and make use of the non-uniform Haar transform. To illustrate this two numerical examples are considered [18].

Example 3.5: Locally disturbed vibrations

Consider the differential equation

$$y'' + k^2 y = f, \; x \in [0, 0.5], \tag{3.38}$$

where

$$f = \begin{cases} d \sin(\alpha x) & \text{for } x \in [x_*, x_{**}], \\ 0 & \text{elsewhere.} \end{cases} \tag{3.39}$$

If x is interpreted as time, then (3.38) and (3.39) presents a locally disturbed sinusoidal motion. Equation (3.38) can be integrated analytically but for conciseness sake we do not show these results here.

Subsequent calculations were carried out for the parameters $k = \pi$, $d = 400$, $\alpha = 40\pi$, $x_* = 0.2$, $x_{**} = 0.25$ and for the initial conditions $y(0) = 1$, $y'(0) = 0$. Let us divide the interval $[0, 0.5]$ into three parts $[0, 0.2]$, $[0.2, 0.25]$, $[0.25, 0.5]$; with the length of the subintervals Δx_1, Δx_2, Δx_3 which are assumed to be constants. It is reasonable to choose the stepsizes so that $\Delta x_1 \approx \Delta x_3$, $\Delta x_1 \gg \Delta x_2$. If we take $\Delta x_1 = 0.4/M$, $\Delta x_2 = 0.05/M$, $\Delta x_3 = 0.5/M$ then to the interval $[0, 0.2]$ belong $M/2$ collocation points, to $[0.2, 0.25]$ M points and to $[0.25, 0.5]$ also $M/2$ points. The coordinates of the grid points are

$$\begin{aligned} &x(l) = l\Delta x_1 \text{ for } l = 1, \ldots, M/2, \\ &x(l + M/2) = x_* + l\Delta x_2 \text{ for } l = 1, \ldots, M, \\ &x(l + 3M/2) = x_{**} + l\Delta x_3 \text{ for } l = 1, \ldots, M/2. \end{aligned} \tag{3.40}$$

We satisfy (3.38) in the collocation points x_l. If Y is a row vector for which $Y(l) = y[x_c(l)]$ then the solution of (3.38) can be sought in the matrix form

$$Y'' = aH. \tag{3.41}$$

By integrating this equation twice we obtain

$$Y' = aP_1 + y'(0)E, \quad Y = aP_2 + y'(0)X + y(0)E. \tag{3.42}$$

Here X_c is a row vector with the components $x_c(l)$ and E is a $2M$-dimensional unit vector. Replacing (3.41) and (3.42) into (3.39) we get

Table 3.3 Error estimates for Eq. (3.38)

J	$2M$	Uniform Haar		Nonuniform Haar	
		δ_{max}	δ_n	δ_{max}	δ_n
3	16	1.59	2.0E-1	4.5E-3	6.9E-4
4	32	0.36	3.0E-2	1.1E-3	1.2E-4
5	64	0.37	2.2E-2	2.8E-4	2.1E-5
6	128	0.34	1.4E-2	6.9E-5	3.7E-6

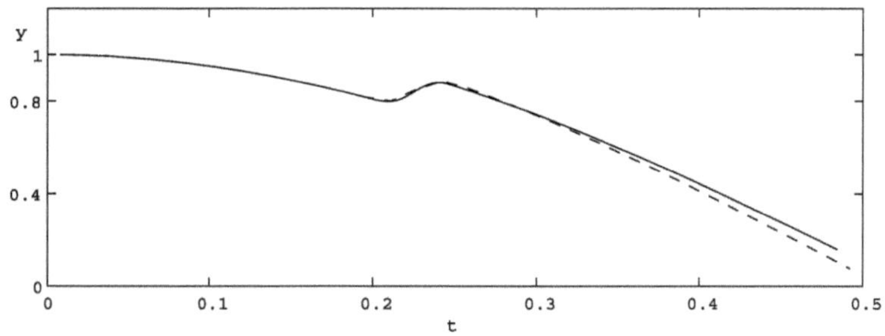

Fig. 3.3 Solution of (3.39) for $J = 5$. Exact solution, uniform wavelet solution [19]

$$a(H + k^2 P_2) = F - k^2[y'(0)X_l + y(0)E], \tag{3.43}$$

where $F(l) = f[x_c(l)]$. Again the vector a is calculated from (3.43); the solution of the problem is found from (3.42). Results of the computer simulation are depicted in Table 3.3 and in Fig. 3.3.

It follows from Table 3.3 and Fig. 3.3 that in case of variable stepsize, the accuracy of the results is very good—already in the case of 16 points (i.e. for $J = 3$) our results visually do not differ from the exact ones. For uniform Haar solution, the error δ_{max} is quite significant, it practically does not become smaller if we increase the number of collocation points to 128. As to the estimate δ_n then it is ~1000 times greater than for nonuniform Haar solution. From here the conclusion, that the uniform Haar method for the solution of this problem is inappropriate, can be made.

Example 3.6: Consider the boundary layer problem

$$-\varepsilon y'' + y' = 0,\ x \in [0, 1],\ y(0) = y(1) = 0, \tag{3.44}$$

where $\varepsilon > 0$ is a small parameter. The exact solution of (3.44) is

$$y_{ex} = x - \frac{e^{x/\varepsilon} - 1}{e^{1/\varepsilon} - 1} \tag{3.45}$$

If $\varepsilon \ll 1$ a boundary layer develops near the edge $x = 1$ (Fig. 3.4). This fact puts an extra requirement to the computational algorithm and therefore (3.44) is a good test problem. It was solved by Bertoluzza [2] who made use of the Deslaurier-Dubuc

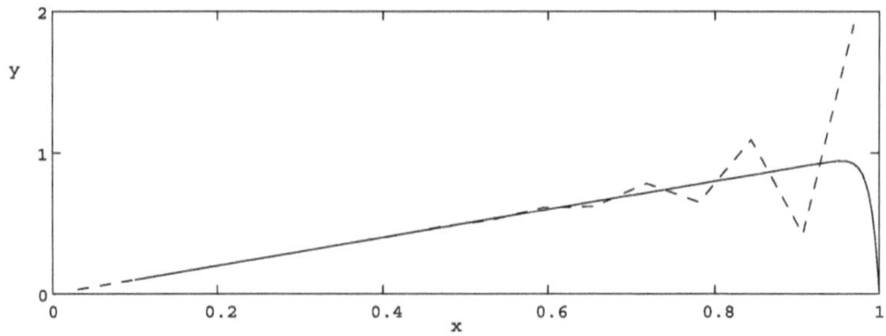

Fig. 3.4 Solution of (3.45) for $\varepsilon = 0.01$; $J = 3$, exact solution and wavelet solution for $q = 0.8$, wavelet solution for $q = 1$ [19]

interpolating wavelets. The Haar wavelet method in connection with the segmentation technique was applied by Lepik [17].

Now let us put together a Haar wavelet solution with the variable stepsize. We shall assume that the lengths of the subintervals fulfill the condition $\Delta x_l = q\Delta x_{l-1}$, where $q < 1$ and $l = 1, \ldots, 2M$.

By summing up all the lengths of the subintervals we obtain

$$\Delta x_1(1 + q + q^2 + \cdots + q^{2M-1}) = 1. \tag{3.46}$$

Since the sum of this geometric series is

$$1 + q + q^2 + \cdots + q^{2M-1} = \frac{1 - q^{2M}}{1 - q} \tag{3.47}$$

we obtain

$$\Delta x_1 = \frac{1 - q}{1 - q^{2M}}. \tag{3.48}$$

The grid points are

$$\tilde{x}_l = \frac{1 - q^l}{1 - q^{2M}}, \quad l = 0, 1, \ldots, 2M. \tag{3.49}$$

The collocation points are calculated according to (2.9).

The solution of the problem (3.44) is sought in the matrix form (3.41). We have to satisfy the boundary condition $y(1) = 0$. According to (3.42) we have

$$y_0' = -aP_2 \mid_{x=1} . \tag{3.50}$$

Let us introduce the row vector K with the components

$$K_i = \frac{1}{2}\left\{[1 - \xi_1(i)]^2 - 2[1 - \xi_2(i)]^2 + [1 - \xi_3(i)]^2\right\}. \tag{3.51}$$

Table 3.4 Error estimates for Eq. (3.44), $\varepsilon = 0.01$

J	$2M$	$q = 1.0$		$q = 0.9$		$q = 0.8$	
		δ_{max}	δ_n	δ_{max}	δ_n	δ_{max}	δ_n
3	16	*	*	0.19	1.2E-2	1.9E-2	1.8E-3
4	32	0.30	9.2E-3	5.8E-3	3.8E-4	2.1E-3	1.4E-4
5	64	0.068	1.3E-3	4.7E-4	2.3E-5	1.9E-3	6.3E-5

Table 3.5 Error estimates for Eq. (3.44), $\varepsilon = 0.001$

J	$2M$	$q = 1.0$		$q = 0.9$		$q = 0.8$	
		δ_{max}	δ_n	δ_{max}	δ_n	δ_{max}	δ_n
3	16	* *	*	*	*	*	*
4	32	* *	*	*	*	4.8E-3	2.6E-4
5	64	*	*	1.3E-3	5.8E-5	1.8E-3	6.3E-5

Calculating $P_2 \mid_{x=1}$ from (2.6) we find $y_0' = -aK^T$, where the index T denotes transposition. Substitution this result into (3.42) gives

$$Y' = a(P_1 - K^T \otimes E), \quad Y = a(P_2 + K^T \otimes X_l), \tag{3.52}$$

where the symbol $\otimes$ denotes the Kronecker tensor product. In view of (3.42) and (3.52) the equation (3.43) obtains the form

$$a(-\varepsilon H + P_1 - K^T \otimes E) = E, \tag{3.53}$$

from which the wavelet coefficients are calculated. The second equation of (3.42) gives the solution of the problem. Calculations are carried out for $\varepsilon = 0.01$ and $\varepsilon = 0.001$; for the parameter the values $q = 1$, $q = 0.9$ and $q = 0.8$ were taken. The error estimates are shown in Tables 3.4 and 3.5. In some cases the solution is unstable—oscillatory; such cases are marked in these tables with an asterisk.

From the data of Tables 3.4 and 3.5 the preference of the Haar method with variable stepsize again becomes evident. To demonstrate this more clearly let us consider the solution for the case $\varepsilon = 0.01$, $J = 3$ (Fig. 3.4). For $q = 0.8$ the wavelet solution visually coincides with the exact solution, while for $q = 1$ (uniform Haar method) the solution is oscillating. If $\varepsilon = 0.001$ the solution for $q = 1$ remains unstable in spite of the increasing number of collocation points.

3.5 Nonlinear Equations

The n-th order nonlinear ODE has the form

$$F(x, y, y', y'', \ldots, y^{(n)}) = 0, \quad x \in [A, B] \tag{3.54}$$

where F is a nonlinear function. We assume again

$$y^{(n)}(x) = \sum_{i=1}^{2M} a_i h_i(x), \tag{3.55}$$

and integrate this equation n-times. These results are discretized in the collocation points and replaced into (3.54). By doing this we get a system of nonlinear equations

$$\phi_l(a_1, a_2, \dots, a_{2M}) = 0, \quad l = 1, 2, \dots, 2M, \tag{3.56}$$

from which the wavelet coefficients a_i can be calculated.

For solving the system of equations (3.56) the Newton method is applied. If the upper index (ν) denotes ν-th iterate, then according to Newton method we have

$$\sum_{i=1}^{2M} (\frac{\partial \phi_l}{\partial a_i})^{(\nu)} \Delta a_i^{(\nu)} = -\phi_l^{(\nu)}, \quad \nu = 0, 1, 2, \dots, \tag{3.57}$$

where $\Delta a_i^{(\nu)} = a_i^{(\nu+1)} - a_i^{(\nu)}$. It is convenient to write this equation in the matrix form. By introducing the row vectors $a^{(\nu)} = (a_i^{(\nu)})$, $\Phi^{(\nu)} = (\phi_l^{(\nu)})$ and the Jacobi matrix $S^{(\nu)} = (\partial \phi_l \partial a_i)_\nu$ we obtain

$$S^{(\nu)} \Delta a^{(\nu)} = -\Phi^{(\nu)}. \tag{3.58}$$

The solution of this equation is

$$\Delta a^{(\nu)} = -\Phi^{(\nu)} / S^{(\nu)}. \tag{3.59}$$

The corrected value for the wavelet coefficients is calculated from the condition

$$\tilde{a}^{(\nu)} = a^{(\nu)} + \lambda \Delta a^{(\nu)}, \quad 0 \le \lambda \le 1, \tag{3.60}$$

where λ is selected to guarantee a decrease of the norm $\| \Phi^{(\nu)} \|$ (in the case of the exact solution we have $\| \Phi \| = 0$).

By applying the Newton method two problems appear

(i) Convergence of the sequence of Newton iterates. This problem is not discussed in the present book and we recommend here to consult a textbook (e.g. [16]).
(ii) The other problem is what to take for the initial iterate $a^{(0)}$. It is a difficult problem, since the initial values $a^{(0)}$ should satisfy the system (3.56) ($2M$ nonlinear algebraic equations!). A random guess for $a^{(0)}$ as a rule does not work.

We recommend for the solution of this problem the following way [22]. Let us assume that the problem is solved for the level $\nu = J - 1$, to which correspond $M = 2^J$ collocation points. The wavelet coefficients for this approximation are labeled as $a_i^{(\nu)}, i = 1, 2, \ldots, M$. Next the value of J is increased by one, thus the number of collocation points is doubled. The wavelet coefficients at the new level are estimated as

$$\hat{a}_i^{(\nu+1)} = \begin{cases} a_i^{(\nu)} & \text{for } i = 1, 2, \ldots, M\,, \\ 0 & \text{for } i = M+1, \ldots, 2M\,, \end{cases} \tag{3.61}$$

The estimates $\hat{y}^{(\nu+1)}$, $\hat{(y')}^{(\nu+1)}$ are calculated according to (3.4) and (3.5). The values $\hat{a}_i^{(\nu+1)}$ are corrected by (3.60) and we obtain

$$a_i^{(\nu+1)} = \hat{a}_i^{(\nu+1)} + \lambda \Delta a_i^{(\nu+1)}, \quad i = 1, 2, \ldots, 2M. \tag{3.62}$$

The corrected values for $y^{(\nu+1)}$, $(y')^{(\nu+1)}$ are calculated again from (3.4) and (3.5). The cycle is repeated until the necessary exactness of the results is obtained. Assumption (3.61) is motivated by the fact that higher coefficients of the sequence a_i are usually quite small.

The question of how to start this procedure remains. We recommend to solve (3.56) by some numerical method for a small number of collocation points (e.g. 2 or 4 points) and take the results for the initial iterate $a^{(0)}$.

Example 3.7: Solve the boundary value problem for

$$\begin{aligned} &2y'y''' - 3y''^2 = 0, \quad x \in [0, 2]\,, \\ &y(0) = -1, \;\; y'(0.5) = 4/3, \;\; y(2) = 1. \end{aligned} \tag{3.63}$$

This is a more general boundary value problem to be compared with the usual two-point boundary problems, since here also a boundary condition at the inner boundary $x = 0.5$ exists.

In view of (3.55) we have

$$\begin{aligned} y'''(x) &= \sum_{i=1}^{2M} a_i h_i(x), \\ y''(x) &= \sum_{i=1}^{2M} a_i p_{1,i}(x) + y''(0), \\ y'(x) &= \sum_{i=1}^{2M} a_i p_{2,i}(x) + y''(0)x + y'(0), \\ y(x) &= \sum_{i=1}^{2M} a_i p_{3,i}(x) + y''(0)x^2 + y'(0)x + y(0). \end{aligned} \tag{3.64}$$

Satisfying the boundary conditions $y'(0.5) = 4/3$, $y(2) = 1$ we obtain

$$
\begin{aligned}
\frac{4}{3} &= \sum_{i=1}^{2M} a_i p_{2,i}(0.5) + \frac{1}{2} y''(0)x + y'(0), \\
1 &= \sum_{i=1}^{2M} a_i p_{3,i}(2) + 2y''(0) + 2y'(0) - 1.
\end{aligned} \tag{3.65}
$$

Evaluating from this system $y'(0)$, $y''(0)$, we get

$$
\begin{aligned}
y'(0) &= \frac{5}{3} + \frac{1}{2} \sum_{i=1}^{2M} a_i q_{1,i}, \\
y''(0) &= -\frac{2}{3} - \frac{1}{2} \sum_{i=1}^{2M} a_i q_{2,i},
\end{aligned} \tag{3.66}
$$

where

$$
\begin{aligned}
q_{1,i} &= p_{3,i}(2) - 4p_{1,i}(0.5), \\
q_{2,i} &= p_{3,i}(2) - 2p_{2,i}(0.5).
\end{aligned} \tag{3.67}
$$

Equation (3.56) obtains the form

$$
\phi_l = 2y'(x_l)y''(x_l) - 3\left[y''(x_l)\right]^2, \tag{3.68}
$$

where the quantities $y'(x_l)$, $y''(x_l)$, $y'''(x_l)$ must be replaced according to formulas (3.64).

The elements of the matrix S are

$$
S(i,l) = 2\left[y'(x_l)\frac{\partial y'''(x_l)}{\partial a_i} + y'''(x_l)\frac{\partial y'(x_l)}{\partial a_i} - 3y''(x_l)\frac{\partial y''(x_l)}{\partial a_i}\right]. \tag{3.69}
$$

In view of (3.64)–(3.66) we have

$$
\begin{aligned}
\frac{\partial y'''(x_l)}{\partial a_i} &= h_i(x_l), \\
\frac{\partial y''(x_l)}{\partial a_i} &= p_{1,i}(x_l) - q_{2,i}, \\
\frac{\partial y'(x_l)}{\partial a_i} &= p_{2,i}(x_l) - q_{2,i}(x_l) + \tfrac{1}{2} q_{1,i}.
\end{aligned} \tag{3.70}
$$

Next, some results of the computer simulation are presented. For beginning two collocation points $x_1 = 0.5, x_2 = 1.5$ are taken. Computations showed that $\| \phi \|$ is minimal for $a_1 = 0.7, a_2 = 0.6$.

At the level $J = 1$ (four collocation points) the estimate (3.61) is $\tilde{a}_i = [0.7, 0.6, 0, 0]$. Correcting it with the Newton's method we find that $\| \phi \| = min$ for $\lambda = 0.14$ and $a = (0.98, 0.83, 0.40, 0.02)$; the error estimates were $\delta_1 =$

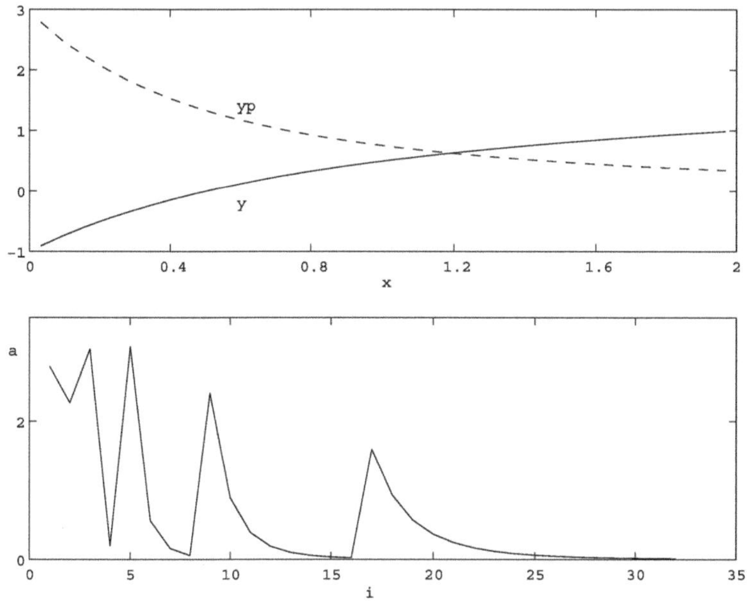

Fig. 3.5 Solution of (3.58) for $J = 4$. **a** $y = y(x)$, $y' = y'(x)$; **b** wavelet coefficients [21]

$2.61E - 2, \sigma_1 = 6.5E - 3$. This process is continued until the necessary exactness is achieved.

The error estimates for the subsequent levels of resolution were

$$\begin{aligned} \delta_2 &= 1.5E - 2, \ \sigma_2 = 1.8E - 3, \\ \delta_3 &= 2.1E - 3, \ \sigma_3 = 1.3E - 4, \\ \delta_4 &= 1.8E - 3, \ \sigma_4 = 5.6E - 5. \end{aligned} \tag{3.71}$$

Numerical results for $J = 4$ (32 collocation points) are plotted in Fig. 3.5.

3.6 Method of Segmentation

If we want to raise the exactness of the results obtained by the Haar wavelet method we must increase the level of resolution J. This way has, nevertheless, some limitations. As it follows from (3.59) for solving ODE-s we must invert the matrix S, but by increasing J the determinant of this matrix decreases and for bigger J can acquire very small values. This means that the matrix S is nearly singular and its inversion brings a considerable loss of exactness. To avoid such a situation the method of segmentation was worked out [17].

Let us explain the idea of this method for solving a second order ODE

$$\frac{d^2y}{dx^2} = F(x, y, \frac{dy}{dx}), \quad x \in [A, B]. \tag{3.72}$$

Instead of (3.72) a system of first order equations

$$\frac{dy}{dx} = z, \quad \frac{dz}{dy} = F(x, y, z) \tag{3.73}$$

is solved.

The interval of integration $x \in [A, B]$ is divided into N segments, the length of the n-th segment is denoted $d^{(n)}$. If $x^{(n)}$ is the coordinate of the n-th grid point, then

$$x^{(n+1)} = x^{(n)} + d^{(n)}, \quad n = 1, 2, \ldots, N, \tag{3.74}$$

with $x^{(1)} = A, \; x^{(n+1)} = B$. Consider the n-th segment. It is convenient to introduce the local coordinate

$$X = \frac{x - x^{(n)}}{d^{(n)}} \tag{3.75}$$

and define $2M$ collocation points

$$X_c(l) = \frac{1}{2M}(l - \frac{1}{2}), \quad l = 1, 2, \ldots, 2M. \tag{3.76}$$

The solution of (3.73) is sought in the form

$$\frac{dy}{dx} = aH, \quad \frac{dz}{dx} = bH, \tag{3.77}$$

where a and b are the wavelet coefficients vectors. By integrating (3.72) we get

$$\begin{aligned} y &= aP_1 + y^{(n)}E, \\ z &= bP_1 + z^{(n)}E. \end{aligned} \tag{3.78}$$

The integral matrix P_1 is calculated from (2.6) and (2.7) assuming $A = 0, B = 1$; E denotes a unit vector $E = [1, 1, \ldots, 1]$.

The results (3.76) are replaced into the (3.73) and we get for calculating the coefficients a, b the system

$$\begin{aligned} aH &= a^{(n)}(bP_1 + z^{(n)}E), \\ bH &= d^{(n)}(F(x(n) + d^{(n)}X_c, \; aP_1 + y^{(n)}E, \; bP_1 + z^{(n)}E)). \end{aligned} \tag{3.79}$$

Solving (3.79) we get the vectors a, b. Replacing the results into (3.78) we can evaluate the functions $y \in \left[y^{(n)}, y^{(n+1)}\right], z \in \left[z^{(n)}, z^{(n+1)}\right]$. To proceed to the next segment $y^{(n+1)}, z^{(n+1)}$ must be computed. Taking into account the fact that $P_1 \mid_{x=1} = [1, 0, \ldots, 0]$ we find from (3.78)

$$y^{(n+1)} = a_1 + y^{(n)}E,\\ z^{(n+1)} = b_1 + z^{(n)}E. \tag{3.80}$$

Here a_1, b_1 denote the first wavelet coefficients for the n-th segment.

The values $y^{(n+1)}$, $z^{(n+1)}$ are the starting values for the $(n+1)$-th segment.

If in each segment only one collocation point is located, then we get the *piecewise constant approximation* (PCA). An alternative to the segmentation method is the non-uniform Haar method, which was discussed in Sect. 3.4.

With the purpose to estimate the good features and the shortcomings of different variants of solution, computer simulation for five problems was carried out [17]. The following programs were tested:

(i) CHM—conventional Haar method,
(ii) SM2—segmentation method with two collocation points in each segment,
(iii) SM4—segmentation method with four collocation points in each segment,
(iiii) PCA—piecewise constant approximation. For details of these solutions consult the paper [17].

A disadvantage of the PCA is that considerably more grid points are needed as in the case of Haar wavelet methods (usually the order of these values is from $N = 100$ to $N = 1000$). Otherwise PCA is very convenient for solving initial value problems for ODEs, since we do not need to calculate the coefficient matrix and to invert it. The situation is different for solving high-order boundary value problems; in the case of PCA we have to solve a large system nonlinear equations (N equations). Here the method based on the Haar wavelet method and discussed in Sect. 3.5 is favourable since the boundary conditions are incorporated into the system of equations and satisfied automatically.

The main advantages of the CHM are sparse representations, fast transformation and the possible implementation of fast algorithms and universal subprograms.

3.7 Related Papers

Chen and Hsiao [6] were evidently the first who applied Haar wavelets for solving ODEs. They investigated the system of equations

$$\dot{x}(t) = Ax(t) + Bu(t),\\ y(t) = Cx(t) + Du(t), \ \ x(0) = x_0, \tag{3.81}$$

Here $u(t)$ denotes a prescribed vector, which is square integrable in the interval of integration; A, B, C, D, x_0 are given constants.

According to the Chen-Hsiao's idea into the Haar series is expanded the variable $\dot{x}(t)$ instead of $x(t)$ itself:

$$\dot{x}(t) = FH(t). \tag{3.82}$$

Integration of this equation gives

$$x(t) = FPH(t) + x_0. \tag{3.83}$$

Here P is the *Haar operational matrix of integration.* It is calculated as it was shown in Sect. 2.6.

Expanding into the Haar series also the function $u(t)$ we find

$$u(t) = GH(t). \tag{3.84}$$

Here F and G denote the coefficient vectors.

Replacing (3.82)–(3.84) into (3.81) and satisfying this system in the collocation points we get a linear system for calculating F and G. This approach has been followed in many papers. First we discuss here the papers by Hsiao and his co-workers in which time varying systems were investigated [13–15].

A typical set up of the problem is: integrate the system

$$E(t)\dot{x}(t) = f(t, x(t), u(t)), \quad x(0) = x_0, \tag{3.85}$$

where $E(t)$ is a singular matrix $E(t) \in R^{n\times n}$, the state variable is $x(t) \in R^n$, control variable $u(t) \in R^q$, $f \in R^n$ denotes a nonlinear function. The response $x(t)$, $t \in (0, 1)$ is required to be found. For solving this problem two new matrices—the *Haar product matrix* and *coefficient matrix*—are introduced.

A *single term Haar wavelet transform* (STHWT) was introduced by Hsiao and Wang [14]. Consider a system of differential equations

$$\dot{x}(t) = F(t, x(t), u(t)), \; x(0) = x_0, \; t \in [0, T]. \tag{3.86}$$

Divide the time interval $(0, T)$ into N parts of equal length $\Delta t = T/N$. Examine the i-th segment (t_i, t_{i+1}) and normalize this interval with

$$\tau = i(t - t_i)/\Delta t. \tag{3.87}$$

Equation (3.86) obtains now the form

$$\dot{x}(\tau) = \tilde{F}(\tau, x(\tau), u(\tau)), \; \tau \in [0, 1]. \tag{3.88}$$

Let us express $\dot{x}(\tau), x(\tau), u(\tau)$ in single term Haar series as

$$\dot{x}_i(\tau) = a_i h_0(\tau), \; x_i(\tau) = b_i h_0(\tau), \; u_i(\tau) = r_i h_0(\tau). \tag{3.89}$$

Since for the first segment $P = 1/2$ the following recursive relationship is obtained [14]

$$b_i = \frac{1}{r}a_i + x_i(j-1),\ x_i(j) = a_i x_i(j-1) = 2b_i - x_i(j-1), \tag{3.90}$$

where $i, j = 0, 1, \ldots, N$.

This approach has been applied in several papers from which we note here the paper by Bujurke et al. [3]. Here the Duffing equation

$$\ddot{x} + A\dot{x} + \omega^2 h(x) = Q(x) \tag{3.91}$$

is solved.

The STHWT is very close to the PCA (piecewise constant approximation method), which was discussed in Sect. 3.6.

In all these papers the solution is based on the evaluation of the operational matrix of integration P (henceforth we call this approach in short the *operation matrix method*). In the case of an alternative method, which is applied in the present book, no operation matrices are introduced and the Haar wavelets are integrated directly arbitrary times (consult Sect. 2.1). This approach we call the *direct integration method*. Now let us consider some papers where the latter method is used. Here mainly boundary value problems of different ODEs are solved.

Second order boundary value problems by Haar collocation method were investigated by Siraj-ul-Islam et al. [27] and by Fazal-i-Haq et al. [10]. Higher order boundary value and eigenvalue problems are solved in a paper by Chang and Piau [5], Shi and Cao [26], Fazal-i-Haq and Ali [9]. In these papers, efficiency and simple applicability of the Haar method for a variety of boundary conditions are stated.

If we compare the operation matrix method and direct integration method, then the second method is evidently more simple (there is no need to calculate the auxiliary matrices, as the coefficient and Haar product matrix). Besides, if we make use of the programs from Chap. 2 the calculation of the integral matrices of arbitrary order is very simple and not very computer time consuming.

Now let us discuss some papers in which other wavelet families are applied. Cheng and Hsu [7] analysed bilinear systems via block-pulse functions.

Legendre wavelet was used for solving different types of ODEs. Singular ODEs were solved by Mohammadi and Hosseini [23]. Lane-Emden type equation

$$y''(x) + \frac{\alpha}{x}y'(x) + f(x, y) = g(x),\ \alpha \geq 0 \tag{3.92}$$

was discussed by Yousefi [29] and also by Pandey et al. [25]. Venkatesh et al. [28] considered the Bratu-type equation

$$y''(x) + \lambda e^{y(x)} = 0, \quad 0 < x < 1. \tag{3.93}$$

In several papers for solving ODEs, the Chebyshev wavelets have been applied. Babolian and Fattahzadeh [1] put together the operation matrix of integration and the product operation matrix for these wavelets. Doha et al. [8] investigated high-order

ODEs. Linear and nonlinear singular boundary problems were treated by Nasab et al. [24]. Numerical solution of time-varying delay systems was presented by Ghasemi and Kajani [11]. In connection with these papers we would like to cite Hsiao and Wang [14], who write "As for Laguerre polynomials, Chebyshev polynomials and Legendre polynomials, the calculation processes are usually too tedious, although some recursive formulae are available. These polynomials are in no way able to compare with Haar wavelets expansion with respect to computation time and data storage requirements".

Cattani [4] applied the Haar wavelet based technique for investigating sharp jumps (spikes) of time-series. The N-length discrete time series is segmented into σ segments, for each segment the Haar wavelet coefficients are calculated: they depend on three parameters: scaling and translation parameter, the number of the segments. The spikes are classified according to the relations existing between the wavelet coefficients.

Computation of eigenvalues and solutions of the Sturm-Liouville equation

$$(p(x)y')' + q(x)y = \lambda r(x)y \tag{3.94}$$

were discussed by Bujurke et al. [3]; to avoid the calculation of the second derivative y'' the problem is reduced to solution of a variational problem.

In the paper [12] by Guf and Jiang the Haar wavelets operation matrix of integration is derived.

References

1. Babolian, E., Fattahzadeh, F.: Numerical solution of differential equations by using Chebyshev wavelet operational matrix of integration. Appl. Math. Comput. **188**, 417–426 (2007)
2. Bertoluzza, S.: An adaptive collocation method based on interpolating wavelets. In: Dahmen, W., Kurdila, A.J., Ostwalds, F. (eds.) Multiscale Wavelet Methods for PDE, pp. 7630–7636. Academic Press, New York (1997)
3. Bujurke, N., Shiralashetti, S., Salimath, C.: An application of single-term Haar wavelet series in the solution of nonlinear oscillator equations. J. Comput. Appl. Math. **227**, 234–244 (2009)
4. Cattani, C.: Haar wavelet-based technique for sharp jumps classification. Math. Comput. Model. **39**, 255–278 (2004)
5. Chang, P., Piau, P.: Haar wavelet matrices designation in numerical solution of ordinary differential equations. Int. J. Appl. Math. **38**, 164–168 (2008)
6. Chen, C., Hsiao, C.: Haar wavelet method for solving lumped and distributed-parameter systems. IEEE Proc. Control Theory Appl. **144**, 87–94 (1997)
7. Cheng, B., Hsu, N.: Analysis and parameter estimation of bilinear systems via Block-Pulse functions. Int. J. Control **36**, 53–65 (1982)
8. Doha, E., Abd-Elhameed, W., Bassuony, M.: New algorithms for solving high even-order differential equations using third and fourth Chebyshev-Galerkin methods. J. Comput. Phys. **236**, 563–579 (2013)
9. Fazal-i-Haq, Ali, A.: Numerical solution of fourth-order boundary-value problems using Haar wavelets. Appl. Math. Sci. **5**, 61–64 (2011)

10. Fazal-i-Haq, Siraj-ul-Islam, Aziz, I.: Numerical solution of singularly perturbed two-points BVPs using nonuniform Haar wavelets. Int. J. Comput. Methods Eng. Sci. Mech. **12**, 168–175 (2011)
11. Ghasemi, M., Kajani, M.: Numerical solution of time-varying delay systems by Chebyshev wavelets. Appl. Math. Model. **35**, 5235–5244 (2011)
12. Guf, J., Jiang, W.: The Haar wavelets operational matrix of integration. Int. J. Syst. Sci. **27**, 623–628 (1996)
13. Hsiao, C., Wang, W.: State analysis and optimal control of linear time-varying systems via Haar wavelets. Optimal Control Appl. Methods **19**, 423–433 (1998)
14. Hsiao, C., Wang, W.: State analysis of time-varying singular bilinear systems via Haar wavelets. Math. Comput. Simul. **52**, 11–20 (2000)
15. Hsiao, C., Wu, S.: Numerical solution of time-varying functional differential equations via Haar wavelets. Appl. Math. Comput. **188**, 1049–1058 (2007)
16. Kelley, C.: Solving nonlinear equations with Newton's method. In: Higham, N.J. (ed.) Serial Fundamentals of Algorithms, SIAM, Philadelphia (2003)
17. Lepik, Ü.: Numerical solution of differential equations using Haar wavelets. Math. Comput. Simul. **68**, 127–143 (2005)
18. Lepik, Ü.: Numerical solution of evolution equations by the Haar wavelet method. Appl. Math. Comput. **185**, 695–704 (2007)
19. Lepik, Ü.: Haar wavelet method for solving higher order differential equations. Int. J. Math. Comput. **1**, 84–94 (2008)
20. Lepik, Ü.: Solving differential and integral equations by the Haar wavelet method; revisited. Int. J. Math. Comput. **1**, 43–52 (2008)
21. Lepik, Ü.: Solving integral and differential equations by the aid of nonuniform Haar wavelets. Appl. Math. Comput. **198**, 326–332 (2008)
22. Lepik, Ü., Tamme, E.: Solution of nonlinear Fredholm integral equations via the Haar wavelet method. Proc. Estonian Acad. Sci. Phys. Math. **56**, 17–27 (2007)
23. Mohammadi, F., Hosseini, M.: A new Legendre wavelet operational matrix of derivative and its applications in solving the singular ordinary differential equations. J. Franklin Inst. **348**, 1787–1796 (2011)
24. Nasab, A., Kiliçman, A., Babolian, E., Atabakan, Z.: Wavelet analysis method for solving linear and nonlinear singular boundary value problems. Appl. Math. Model. **37**, 5876–5886 (2013)
25. Pandey, R., Kumar, N., Bhardwaj, A., Dutta, G.: Solution of Lane-Emden type equations using Legendre operational matrix of differentiation. Appl. Math. Comput. **218**, 7629–7637 (2012)
26. Shi, Z., Cao, Y.: Application of Haar wavelet method to eigenvalue problems of high order differential equations. Appl. Math. Model. **9**, 4020–4026 (2012)
27. Siraj-ul-Islam, Aziz, I., Sarler, B.: The numerical solution of second-order boundary value problems by collocation method with the Haar wavelets. Math. Comput. Model. **52**, 1577–1590 (2010)
28. Venkatesh, S., Ayyaswamy, S., Balachandar, S.: The Legendre wavelet method for solving initial value problems of Bratu-type. Comput. Math. Appl. **63**, 1287–1295 (2012)
29. Yousefi, S.: Legendre wavelets method for solving differential equations of Lane-Emden type. Appl. Math. Comput. **181**, 1417–1422 (2006)

Chapter 4
Stiff Equations

4.1 Introduction

In the last time, stiff differential equations have been studied extensively and various methods for their solutions have been proposed. While the intuitive meaning of stiffness is clear to all specialists, a unique mathematical definition is lacking.

As a rule, stiffness occurs in differential equations where there are two or more different time scales of the independent variables on which the dependent variables are changing. In the case of linear problems, the stiffness is caused by eigenvalues of big negative values. To measure the degree of stiffness, the following stiffness ratio $SR = \max |\lambda| / \min |\lambda|$ is introduced: when $SR < 20$ the problem is not stiff, up to $SR = 1000$, the problem is classified as stiff, and when $SR \geq 100000$, the problem is very stiff. Stiff ODEs are called extremely stable if there is at least one eigenvalue with a large negative real part. In the case of nonlinear problems, the problem is more complicated since stiffness is a global problem and cannot be reduced to the solution structure in the neighborhood of single points.

If we apply, for the solution of stiff differential equations, explicit solvers (e.g. MatLab solvers ODE45, ODE23s) the solution may be oscillating or diverge (oscillatory solution is a result of the failure of the solution algorithm). For avoiding these oscillations implicit solvers or explicit solvers with a very small stepsize must be used. All this complicates the numerical work. Besides, small steps introduce too many round-off errors and cause numerical instabilities. Depending on the nature of stiffness, different methods of solution should be used.

A detailed analysis of the methods for solving stiff ordinary differential equations can be found in the text-book by Hairer and Wanner [7], to this book an extensive list of publications is added. We would like to turn attention also to the paper by Enright et al. [4] in which different numerical methods have been tested for solving 25 systems of stiff equations.

In the last time for numerical solution of stiff equations, a variety of methods have been applied. From the different ways of solution, we mention here the Adomian

Ü. Lepik and H. Hein, *Haar Wavelets*, Mathematical Engineering,
DOI: 10.1007/978-3-319-04295-4_4,

decomposition method [13], power series method [8] and a comparative study of different MatLab ODE solvers [1].

The main goal of this Chapter is to demonstrate that the Haar wavelet approach can successfully compete with the other methods. For this purpose a number of test problems about linear and nonlinear stiff equations are solved by the Haar method. A special case of the stiff equations form the singular perturbation problems is discussed in Sect. 4.5. The following presentation is based on the paper by Lepik [12].

4.2 Linear Problems

Consider a n-th order ODE

$$y^{(n)}(x) = f(x, y, y', \ldots, y^{(n-1)}), \quad x \in [A, B], \tag{4.1}$$

where f is a linear function. The solution of (4.1) is sought in the matrix form

$$y^{(n)} = aH, \tag{4.2}$$

where a is the vector of wavelet coefficients $a_i, i = 1, 2, \ldots, 2M$.

As shown in Chap. 3 we integrate (4.2) n times and replace the ingredients, which are discretized by the collocation points, into (4.1). In this way we get a system of linear equations for calculating the wavelet coefficients a_i.

Let us pass to some examples.

Example 4.1:

Press et al. [14] investigated the system

$$\begin{aligned} y' &= 998y + 1998z, \quad x \in [0, 1], \\ z' &= -999y - 1999z, \ y(0) = 1, \ z(0) = 0. \end{aligned} \tag{4.3}$$

We reduce (4.3) to the second order ODE

$$y'' + 1001y' + 1000y = 0, \quad x \in [0, 1], \quad y(0) = 1, \quad y'(0) = 0. \tag{4.4}$$

The exact solution of (4.4) is

$$y_{ex} = \frac{1}{999}(1000e^{-x} - e^{-1000x}). \tag{4.5}$$

The eigenvalues of (4.4) are $\lambda_1 = -1, \lambda_2 = -1000$; since $|\lambda_2| >> |\lambda_1|$ the system (4.3) is stiff. It is demonstrated by Press et al. [14] that explicit methods of integration bring to oscillations and the solution is unstable.

Now let us solve (4.4) by the Haar wavelet method. The solution is sought in the form

$$\begin{aligned} y'' &= aH, \\ y' &= aP_1 + y_0' E, \\ y &= aP_1 + y_0' x + y_0 E. \end{aligned} \tag{4.6}$$

Here $E = (1, 1, ..., 1)$ and all quantities are calculated in the collocation points. Replacing (4.6) into (4.4) we get the matrix equation

$$a(H + 1001 P_1 + 1000 P_2) = -1000 E \tag{4.7}$$

from which the wavelet coefficients a_i are calculated. The solution of the problem $y = y(x)$ is found from $(4.6)_3$.

Computer simulation gave the following error estimates

$$\begin{aligned} \delta_3 &= 4.5e-4, \quad \sigma_3 = 2.8e-5, \\ \delta_4 &= 1.7e-4, \quad \sigma_4 = 5.3e-6, \\ \delta_5 &= 4.1e-5, \quad \sigma_5 = 6.5e-7. \end{aligned} \tag{4.8}$$

It follows from here that high precision of the results is guaranteed already for a small number of collocation points (e.g. if $J = 3$ we have only 16 points).

Example 4.2: Mahmood and Casas [13] solved by the Adomian decomposition method the linear initial value problem

$$\begin{aligned} y_1' &= -20y_1 - 0.25y_2 - 19.75y_3, \quad & y_1(0) = 1 \\ y_2' &= 20y_1 - 20.25y_2 + 0.25y_3, \quad & y_2(0) = 0 \\ y_3' &= 20y_1 - 19.75y_2 - 0.25y_3, \quad & y_3(0) = -1 \end{aligned} \tag{4.9}$$

which has an exact solution

$$\begin{aligned} y_{1ex} &= \frac{1}{2}\left[e^{-0,5t} + e^{-20t}(\cos 20t + \sin 20t)\right], \\ y_{2ex} &= \frac{1}{2}\left[e^{-0,5t} - e^{-20t}(\cos 20t - \sin 20t)\right], \\ y_{3ex} &= -\frac{1}{2}\left[e^{-0,5t} + e^{-20t}(\cos 20t - \sin 20t)\right]. \end{aligned} \tag{4.10}$$

The solution (4.10) is plotted in Fig. 4.1.

This problem was solved as well by Guzel and Bayram [6] who applied the power series method. Hojjati et al. [8] solved (4.6) with the aid of a predictor-corrector method, based on backward differentiation.

The wavelet solution is sought in the form

$$\begin{aligned} y_1' &= aH, \; y_1 = aP_1 + E \\ y_2' &= bH, \; y_2 = bP_1 \\ y_3' &= cH, \; y_3 = cP_1 - E\,. \end{aligned} \tag{4.11}$$

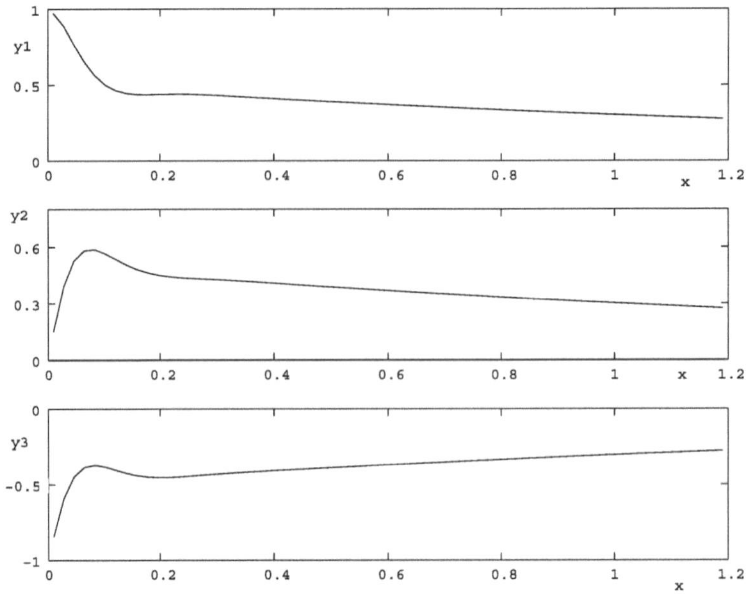

Fig. 4.1 Solution of the system (4.9) [12]

Replacing (4.11) into the system (4.9) we obtain the system

$$\begin{aligned} a(H + 20P_1) + 0.25bP_1 + 19.75cP_1 &= -0.25E \\ -20aP_1 + b(H + 20.25P_1) - 0.25cP_1 &= 19.75E \\ -20aP_1 + 19.75bP_1 + c(H + 0.25P_1) &= 20.25E. \end{aligned} \tag{4.12}$$

Solving (4.12) we calculate the vectors a, b, c; the solution of the problem $y_1(x)$, $y_2(x)$, $y_3(x)$ is obtained from (4.11).

Computations for $J = 5$ (64 collocation points) were carried out and the results were compared with the exact solution (4.10). The error estimates were

$$\begin{aligned} \delta_5(y_1) &= 2.0e-5, \quad \sigma_5(y_1) = 3.1e-7, \\ \delta_5(y_2) &= 4.4e-5, \quad \sigma_5(y_2) = 6.9e-7, \\ \delta_5(y_3) &= 1.6e-3, \quad \sigma_5(y_3) = 2.5e-5. \end{aligned} \tag{4.13}$$

4.3 Nonlinear Equations

Assume that the function in (4.1) is nonlinear. Carrying out the same procedure as in Sect. 4.2 we get for calculation of the wavelet coefficients a_i a nonlinear system of algebraic equations

$$F(a_1, a_2, \ldots, a_{2M}) = 0, \tag{4.14}$$

where F denotes a row vector $F = (f_i),\ i = 1, 2, \ldots, 2M$. This equation is solved by the Newton method as shown in Sect. 3.5.

Let us consider an example.

Example 4.3:

Consider the equation

$$y' = -\frac{y^2}{1+x}, \quad y(0) = 1\,, \quad x \in [0, 200], \tag{4.15}$$

which has the exact solution

$$y = [\ln(1+x) + 1]^{-1}. \tag{4.16}$$

This problem is classified by Hsiao [10] as a problem of "strong instability". Hsiao's solution is based on the single-term Haar wavelet method.

Now let us solve (4.14) by multi-term Haar wavelets. It follows from (4.15) that the solution $y(x)$ in the beginning rapidly decreases and then calms down. Therefore it is reasonable to solve the problem in two stages: first consider the phase of rapid decrease $x \in [0, \delta]$ and after that find a solution for $x \in [\delta, 200]$. In the following calculations the value $\delta = 20$ was chosen. In both regions we take

$$y' = aH, \quad y = aP_1 + y_{in}E. \tag{4.17}$$

Equation (4.14) we solve again by the Newton method. The function $F(x)$ and its gradient get in view of (4.15), (4.17) the form

$$\begin{aligned} F(x) &= aH + y.y/(E+x) \\ \frac{\partial F}{\partial a} &= H + 2y.P_1/(E+x). \end{aligned} \tag{4.18}$$

Here $x = (x_1, x_2, ..., x_{2M})$ and the point denotes element-by-element multiplication, so that

$$(y.^2)_l = y(l)^2; \quad (y.P_1)_{i.l} = y(l)P_1(i,l). \tag{4.19}$$

In the first phase $x \in [0, 20]$ we have $y_{in} = 1$. For the initial solution we take $a^0 \equiv 0$. This value is corrected by the Newton method. By increasing step by step the resolution level J the solution is made more and more exact. Computer simulation for $J = 5$ gave the error estimates $\delta_5 = 6.9e-3,\ \sigma_5 = 6.0e-5$. The value $y(20) = 0.2475$ was calculated, it is taken for the initial value y_{in} of the second stage $x \in [20, 200]$. It is assumed again that $a^0 \equiv 0$ and this value is corrected by the Newton method. Here the number of collocation points may be smaller since already $\delta_3 = 8.5e-3,\ \sigma_3 = 1.6e-4$.

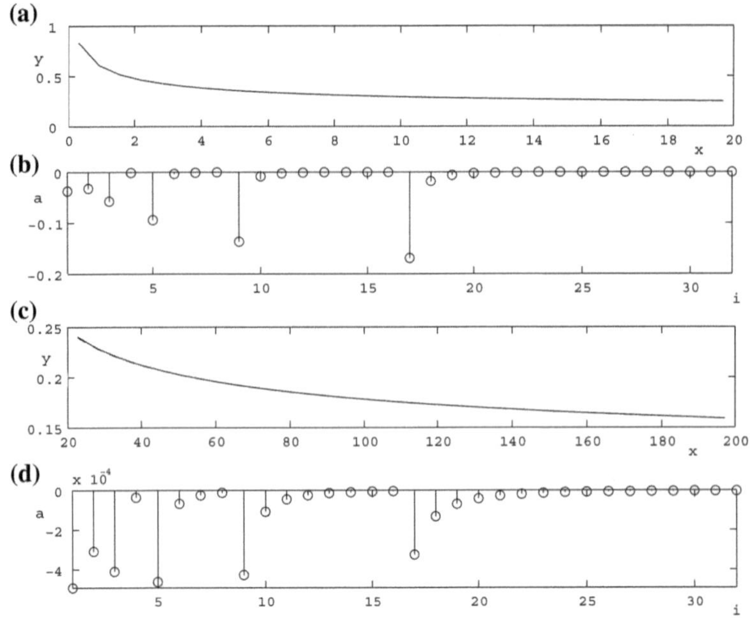

Fig. 4.2 Solution of (4.15) for $J = 4$ **a** first phase $x \in [0, 20]$; **b** wavelet coefficients $a(i)$ for this phase; **c** second phase $x \in [20, 200]$; **d** wavelet coefficients for the second phase [12]

The wavelet solution $y = y(x)$ and the wavelet coefficients $a_i,\ i = 1, 2, \ldots, 2M$ are plotted in Fig. 4.2. It follows from this Figure that the number of significant wavelet coefficients is rather small.

4.4 Robertson's Problem

In 1966 Robertson investigated a chemical system containing fast and slow motions at the same time. By modelling this system he got the system

$$\begin{aligned} y_1' &= -0.04y_1 + 10^4 y_2 y_3, & y_1(0) &= 1, \\ y_2' &= 0.04y_1 - 10^4 y_2 y_3 - 3 \cdot 10^7 y_2^2, & y_2(0) &= 0, \\ y_3' &= 3 \cdot 10^7 y_2^2, & y_3(0) &= 0. \end{aligned} \tag{4.20}$$

The eigenvalues of (4.20) are found in the paper by Abelman and Patidar [1] $\lambda_1 = 0,\ \lambda_2 = -0.405,\ \lambda_3 = -2189.6$; the third one produces the stiffness.

The system (4.20) has been investigated in several papers; it is often taken as a benchmark for numerical methods. A detailed analysis about it can be found in the text-book [7]; it is shown that explicit methods give oscillating solutions (see Fig. 1.3 in the book by Hairer and Wanner). Stability is guaranteed in the case of implicit

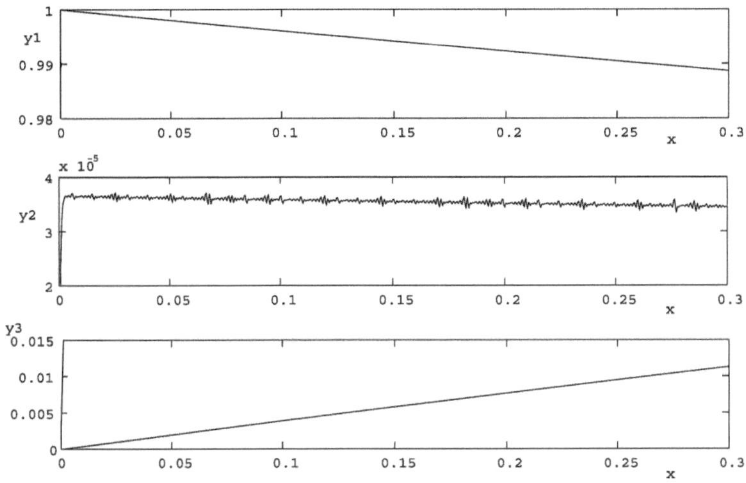

Fig. 4.3 Solution of the Robertson's problem (4.20) obtained with the aid of the program ODE45 [12]

methods (the Runge-Kutta codes DOPRI 5 and RADAU 5 were applied). Abelman and Patidar integrated the system (4.20) by the single term Haar wavelet method and [13] applied for it the Adomian decomposition method. Solution obtained by the ODE45 code is plotted in Fig. 4.3. Small oscillations of the curve y_2 indicate instability of the solution.

Let us pass to the Haar wavelet method. Qualitative analysis of (4.20) indicates that the component y_2 rapidly reaches its maximal value for which $y_2' = 0$ and $3 \cdot 10^7 y_2^2 \approx 0.04$, consequently $\max(y_2) \approx 3.65e - 5$; after that the function y_2 decreases very slowly. Due to this fact it is again expedient to solve (4.20) separately for the regions $x \subset [0, \delta]$ and $x \subset [\delta, x_{max}]$.

The wavelet solution is sought in the form

$$\begin{aligned} y_1' &= aH, & y_2' &= bH, & y_3' &= cH, \\ y_1 &= y_1(0)E + aP_1, & y_2 &= y_2(0)E + bP_1, & y_3 &= y_3(0)E + cP_1. \end{aligned} \tag{4.21}$$

Next the vectors F_1, F_2, F_3 are introduced

$$\begin{aligned} F_1 &= y_1' + 0.04y_1 - 10^4 y_2.y_3, \\ F_2 &= y_2' - 0.04y_1 + 10^4 y_2.y_3 + 3 \cdot 10^7 y_2.^2, \\ F_3 &= y_3' - 3 \cdot 10^9 y_2.^2. \end{aligned} \tag{4.22}$$

On account of (4.21), these vectors are functions of the wavelet coefficients a, b, c. We have to find such values a, b, c for which $F_1 = F_2 = F_3 = 0$. This can be realized according to the following procedure.

(i) We assume that some initial estimates $\tilde{y}_1, \tilde{y}_2, \tilde{y}_3$ are known. From (4.21) we calculate the estimates $\tilde{a}, \tilde{b}, \tilde{c}$ and also $\tilde{y}_1', \tilde{y}_2', \tilde{y}_3'$.

(ii) Evaluate F_2 from $(4.22)_2$.

(iii) We vary only the coefficients b (a and c are fixed) and calculate the gradient

$$\frac{\partial F_2}{\partial b} = H + 10^4 y_3 \cdot P_1 + 6 \cdot 10^7 y_2 \cdot P_1. \tag{4.23}$$

(iv) Making use of the Newton method next approximation for b is calculated:

$$b = \tilde{b} - \lambda F_2 \left(\frac{\partial F_2}{\partial b} \right)^{-1}. \tag{4.24}$$

For the coefficient λ the value which minimizes $\| F_2 \|$ is taken

(v) Calculate from $(4.20)_3$ and (4.21) the corrected values for c, y_3, y_3'.

(vi) Calculate from $(4.20)_1$ and (4.21) the corrected values for a, y_1, y_1'.

(vii) Evaluate $\| F_1 \|_\infty$, $\| F_2 \|_\infty$, $\| F_3 \|_\infty$; if these norms are not sufficiently near to zero repeat the steps (ii)–(vii).

Computer simulation was carried out for $\delta = 0.005$, $x_{max} = 0.3$. In the first phase $x \in [0, \delta]$ the argument x is very small and we can develop the function y_1, y_2, y_3 into power series. Taking into account only the first terms of this series we find

$$\begin{aligned} \tilde{y}_1 &= E - 0.04x, \\ \tilde{y}_2 &= 0.04E - 1.6 \cdot 10^4 x.^3, \\ \tilde{y}_3 &= 1.6 \cdot 10^4 x.^3. \end{aligned} \tag{4.25}$$

According to (4.21) we obtain

$$\tilde{a} = -0.04E/H, \quad \tilde{b} = -4.8 \cdot 10^4 x.^2/H, \quad \tilde{c} = -\tilde{b}.$$

For correcting these values two Newton steps were needed. The final results for $J = 5$ were

$$y_1(\delta) = 0.9998, \quad y_2(\delta) = 3.648e - 5, \quad y_3(\delta) = 1.626e - 5.$$

These values were taken for initial values of the second phase $x \in [0.005, 0.3]$. In this phase the functions y_1, y_3 decrease very slowly, therefore we take

$$\tilde{y}_1 = y_1(\delta)E, \quad \tilde{y}_2 = y_2(\delta)E.$$

Calculating $\tilde{y}_3$ from (4.20) we find

$$\tilde{y}_3 = y_3(\delta)E + 3 \cdot 10^7 y_2(\delta)^2 x.$$

Correcting once again the values by the Newton method we find

$$y_1(0.3) = 0.9888, \quad y_2(0.3) = 3.446e - 5, \quad y_3(0.3) = 1.138e - 4.$$

For comparison the calculations were carried out by the ODE45 code and we get

$$y_1(0.3) = 0.9987, \quad y_2(0.3) = 3.444e - 5, \quad y_3(0.3) = 1.129e - 4.$$

These data are in good accordance with our results.

It is interesting to note that in the second phase, only the first of the wavelet coefficients is significant. To demonstrate this, the first three wavelet coefficients are written down:

$$\begin{aligned} a &: -0.0376, \ -0.0011, \ -0.0006, \ldots \\ 10^5 b &: -0.6851, \ -0.0183, \ -0.0092, \ldots. \\ c &: \ \ \ 0.0377, \ \ \ 0.0011, \ \ \ 0.0006, \ldots \end{aligned}$$

All other coefficients are considerably smaller.

4.5 Singular Perturbation Problems

A singular perturbation problem of Index 1 is defined [7] as

$$\begin{aligned} y' &= f(y, z), \quad y(x_0) = y_0, \\ \varepsilon z' &= g(y, z), \quad z(x_0) = z_0. \end{aligned} \tag{4.26}$$

If ε is small the second equation of (4.26) is stiff. In the case $\varepsilon = 0$ we get a *differential algebraic equation* (here differential equation is combined with an algebraic equation). Equations (4.26) for $\varepsilon = 0$ are called also the *reduced system*, it is usually much easier to analyse than the primary system.

Consider the phase space (y, z). The slope of the phase curves is

$$\frac{dz}{dy} = \frac{g(y, z)}{\varepsilon f(y, z)}. \tag{4.27}$$

We assume that the initial point (y_0, z_0) is not placed on the curve $g(y, z) = 0$ and $\varepsilon << 1$; in this case the phase curve is very steep.

We are mainly interested in the case where the phase point (y, z) moves towards the curve $g(y, z) = 0$ and reaches it at some instant $x = \delta$. Now $z'(\delta) = 0$ and the following motion proceeds along the curve $g(y, z) = 0$. Numerical difficulties can appear in the first phase of motion $x \leq \delta$, since for $\varepsilon << 1$ the second equation of (4.26) is stiff.

Differential-algebraic equations are investigated in many papers from which we cite here [3, 5, 15].

Example 4.4: Consider the van der Pol equation

$$\varepsilon z'' + (z^2 - 1)z' + z = 0, \tag{4.28}$$

which can be rewritten in the form [7]

$$\begin{aligned} y' &= -z, & y(x_0) &= y_0, \\ \varepsilon z' &= y - z^3/3 + z, & z(x_0) &= z_0. \end{aligned} \tag{4.29}$$

Boundary condition for z' can be calculated from $(4.29)_2$:

$$\varepsilon z'(x_0) = \frac{1}{\varepsilon}(y_0 - z_0^3/3 + z_0). \tag{4.30}$$

The reduced problem $\varepsilon = 0$ can be easily solved:

$$z' = \frac{z}{1 - z^2} \tag{4.31}$$

and

$$\ln|z| - 0.5z^2 = x + c. \tag{4.32}$$

Computer simulation was carried out for $\varepsilon = 0.001$, $y_0 = 0.5$, $z_0 = 1$. According to (4.30) we have $z_0' = 7000/6$. Since $z' > 0$ the phase point (y, z) moves toward the phase curve

$$y = z^3/3 - z \tag{4.33}$$

and reaches it at some instant $x = \delta$.

Numerical results obtained by the MatLab program ODE23s and with the aid of the piecewise constant approximation method (PCA) are plotted in Fig. 4.4. It follows from this figure that at this number of calculation points, the PCA solution is unstable.

Now let us present the Haar wavelet solution. Again it is suitable to consider separately two phases of motion: (i) the phase of rapid changes $x \in [0, \delta]$ and (ii) motion along the curve (4.32) for $x \in [\delta, x_{fin}]$. In the following calculations it was taken $\delta = 0.003$, $x_{fin} = 0.02$.

In the first phase the solution is sought in the form

$$\begin{aligned} z'' &= aH, & z' &= aP_1 + z_0'E, \\ z &= aP_2 + z_0'x + z_0E. \end{aligned} \tag{4.34}$$

Next the $2M$-dimensional vector F is defined:

$$F(a) = \varepsilon z'' + (z.^2 - E).z' + z; \tag{4.35}$$

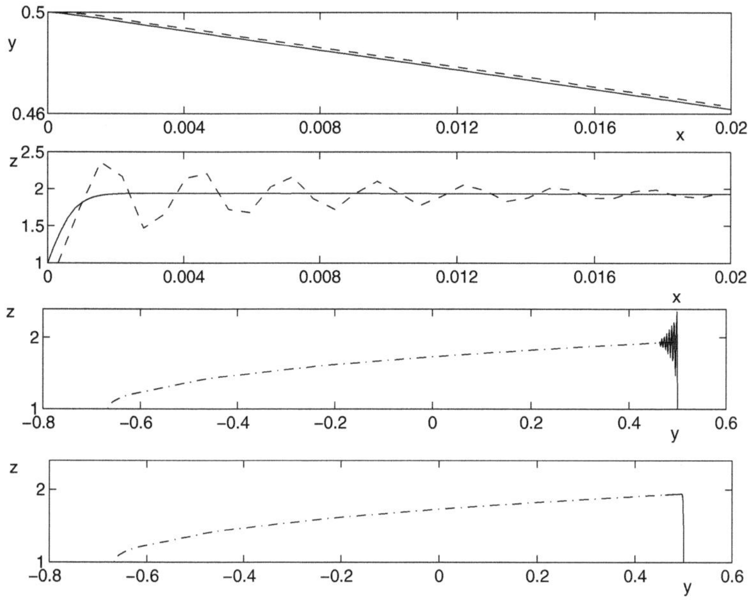

Fig. 4.4 Solution of the van der Pol equation (4.28) **a–b** functions $y(x)$, $z(x)$, dashed lines denote the PCA solutions; **c** phase curve according to the PCA solutions; **d** phase curve for the wavelet solution [12]

its gradient is

$$\frac{\partial F}{\partial a} = \varepsilon H + (z.^2 - E).P_1 + (E + 2z.z').P_2. \tag{4.36}$$

The corrected estimate for wavelet coefficients is

$$a = \tilde{a} - \lambda F \left(\frac{\partial F}{\partial a}\right)^{-1}. \tag{4.37}$$

The coefficient λ is chosen so that $\|F\|_\infty$ would be minimal.

Computer simulation was carried out for $J = 4$, for the initial approximation the value $\tilde{a} \equiv 0$ was taken. For correcting it three Newton steps were made. For the final result we got $\|F\|_\infty = 0.0026$, $y(\delta) = 0.4938$, $z(\delta) = 1.940$, $z'(\delta) = 0.9224$.

In the second phase $x \in [\delta, x_{fin}]$ we assume

$$z' = bH, \quad z = bP_1 + z(\delta)E. \tag{4.38}$$

Now the function

$$G(b) = (z.^2 - E).z' + z \tag{4.39}$$

is to be minimized.

The gradient of (4.38) is

$$\frac{\partial G}{\partial b} = (z.^2 - E).H + (E + 2z.z').P_1. \tag{4.40}$$

Again we assume $\tilde{b} \equiv 0$; after two Newton steps we find $\|G\|_\infty = 8.5e - 5$, $y(0.02) = 0.4629$, $z(0.02) = 1.928$. This is in good accordance with the ODE23s solution which gives $y(0.02) = 0.4618$, $z(0.02) = 1.928$.

As to the wavelet coefficients then, the first coefficient is $b_1 = -0.7058$; all other coefficients are very small (less than 0.27 % from b_1). This is due to the quasi-linearity of the function $z(x)$.

The examples in this Chapter demonstrate that in solving stiff systems, the Haar wavelet method can successfully compete with other methods.

4.6 Related Papers

There are not many papers in which the Haar wavelet method is applied for solving stiff systems. To begin with, we would mark the papers by Hsiao and Wang [9–11]. In these papers linear and nonlinear stiff systems of differential equations are analyzed. The solution based on the single-term Haar wavelet method (STHW), when compared with the other methods, shows that the classical Runge-Kutta method (CRK) can fail completely. Somewhat better results can be obtained for the automatic stepsize Runge-Kutta-Fehlberg method (ASSRKF), but still the results obtained by STHW are more exact. Besides the STHW is very fast. Hsiao and Wang [11] calculated the computation time for these three methods. The results are given e.g. in the case of Example 4.1 the computation time in seconds was 24 for STHW, 3443 for CRK and 315 for ASSRKF.

Bujurke et al. [2] solved numerically some stiff systems from nonlinear dynamics (laser technics, nuclear reactor theory). Here once again the STHW method was applied. Stability of the solutions is also discussed.

References

1. Abelman, S., Patidar, K.: Comparison of some recent numerical methods for initial-value problems for stiff ordinary differential equations. Comput. Math. Appl. **55**, 733–744 (2008)
2. Bujurke, N., Salimath, C., Shiralashetti, S.: Numerical solution of stiff systems from nonlinear dynamics using single-term Haar wavelet series. Nonlinear Dyn. **51**, 595–605 (2008)
3. Çelik, E.: On the numerical solution of differential-algebraic equations with Index - 2. Appl. Math. Comp. **156**, 541–550 (2004)
4. Enright, W., Hull, T., Lindberg, B.: Comparing numerical methods for stiff systems of ODEs. BIT **15**, 10–48 (1975)
5. Gao, J., Jiang, Y.L.: An adaptive wavelet method for nonlinear differential-algebraic equations. Appl. Math. Comp. **189**, 208–220 (2007)

6. Guzel, N., Bayram, M.: On the numerical solution of stiff systems. Appl. Math. Comp. **170**, 230–236 (2005)
7. Hairer, E., Wanner, G.: Solving Ordinary Differential Equations II: Stiff and Differential-Algebraic Problems. Springer, Berlin (1991)
8. Hojjati, G., Ardabili, M.R., Hosseini, S.: A-EBDF: an adaptive method for numerical solution of stiff systems of ODEs. Math. Comput. Simul. **66**, 33–41 (2004)
9. Hsiao, C.: Haar wavelet approach to linear stiff systems. Math. Comput. Simul. **64**, 561–567 (2004)
10. Hsiao, C.: Numerical solution of stiff differential equations via Haar wavelets. Int. J. Comp. Math. **82**, 1117–1123 (2005)
11. Hsiao, C., Wang, W.: Haar wavelet approach to nonlinear stiff systems. Math. Comput. Simul. **57**, 347–353 (2001)
12. Lepik, Ü.: Haar wavelet method for solving stiff differential equations. Math. Model. Anal. **14**, 467–481 (2009)
13. Mahmood, A., Casasús, L., Al-Hayani, W.: The decomposition method for stiff systems of ordinary differential equations. Appl. Math. Comp. **167**, 964–975 (2005)
14. Press, W., Flannery, B., Teukolsky, S., Vetterling, W.: Numerical Recipes. Cambridge University Press, Cambridge (1987)
15. Vijalapura, P., Strain, J., Covindjee, S.: Fractional step methods for index - 1 differential-algebraic equations. J. Comput. Phys. **203**, 305–320 (2005)

Chapter 5
Integral Equations

5.1 Introduction

The integral equations can be classified in the following way

(i) Fredholm equation

$$u(x) = \int_A^B K[x, t, u(t)]dt + f(x), \quad (x, t) \in [A, B] \tag{5.1}$$

(ii) Volterra equation

$$u(x) = \int_A^x K[x, t, u(t)]dt + f(x), \quad (x, t) \in [A, B] \tag{5.2}$$

(iii) Integro-differential equation

$$\alpha u'(x)+\beta u(x) = \int_A^x K[x, t, u(t), u'(t)]dt+f(x), \quad \text{for } u'(A) = u_0', \ x \in [A, B] \tag{5.3}$$

Here α, β, A, B denote prescribed constants, f and K are given functions, the quantity K is called the *kernel*. The problem is to find the unknown function $u(x)$.

If $f(x) = 0$ the integral equation is called *first kind equation*. If $f(x) \neq 0$ we have the *second kind equation*.

If the kernel has the form

$$K[x, t, u(t)] = A(x, t)u(t) \tag{5.4}$$

Ü. Lepik and H. Hein, *Haar Wavelets*, Mathematical Engineering,
DOI: 10.1007/978-3-319-04295-4_5,

the Eqs. (5.1)–(5.3) are *linear integral equations*, if (5.4) does not hold, these equations are nonlinear.

In the case

$$K[x, t, u(t)] = A(x, t)(x - t)^{-\gamma}, \text{ where } \gamma \in (0, 1) \tag{5.5}$$

the corresponding integral equation is *weakly singular*.

Many problems from physics and other disciplines lead to integral equations. Several methods for numerical solution of these equations have been proposed (see e.g. Atkinson [2]). Since 1991 different wavelet methods for solving integral equations have been applied.

This chapter is put together on the basis of the papers [15–20].

For discretizing the equations beside the collocation method, as an alternative the Galerkin method has been applied. It is interesting to compare the solutions obtained by both methods. Such an analysis is carried out in Sects. 5.2–5.3.

5.2 Fredholm Integral Equation

Consider this linear Fredholm integral equation

$$u(x) = \int_A^B K(x, t)u(t)dt + f(x), \quad x \in [A, B]. \tag{5.6}$$

The solution is sought in the form

$$u(x) = \sum_{i=1}^{2M} a_i h_i(x), \tag{5.7}$$

where a_i are the wavelet coefficients and $h_i(x)$ is calculated according to (2.1). Replacing (5.7) into (5.6) we obtain

$$\sum_{i=1}^{2M} a_i h_i(x) - \sum_{i=1}^{2M} a_i G_i(x) = f(x), \tag{5.8}$$

where

$$G_i(x) = \int_A^B K(x, t)h_i(t)dt. \tag{5.9}$$

For calculating the wavelet coefficients a_i we use the two following ways

(i) Collocation method. Satisfying (5.8) only at the collocation points (2.9) we get a system of linear equations

$$\sum_{i=1}^{2M} a_i[h_i(x_l) - G_i(x_l)] = f(x_l),\ l = 1, 2, \dots, 2M. \tag{5.10}$$

The matrix form of this system is

$$a(H - G) = F \tag{5.11}$$

where $G_{il} = G_i(x_l),\ F_l = f(x_l)$.

(ii) Galerkin method. For realizing this approach each term of (5.8) is multiplied by $h_l(x)$ and the result is integrated over $x \in [A, B]$. Due to the orthogonality condition (2.4) we obtain

$$\frac{a_l}{m_l} - \sum_{i=1}^{2M} a_i \Gamma_{il} = \int_A^B f(x) h_l(x) dx. \tag{5.12}$$

Here $l = m + k + 1,\ m = 2^j,\ j = 0, 1, \dots, J,\ k = 0, 1, \dots, m - 1$ and

$$\Gamma_{il} = \int_A^B G_i(x) h_l(x) dx. \tag{5.13}$$

Example 5.1: Let us solve (5.6) for $K(x, t) = x + t,\ f(x) = x^2$. Evaluating the integrals (5.9) we find

$$G_i(x) = \begin{cases} x + 0.5 \text{ for } i = 1, \\ -\frac{1}{4m^2} \text{ for } i > 1. \end{cases} \tag{5.14}$$

If we apply the collocation method, then the vector a can be calculated from the system (5.11). In the case of the Galerkin method, we shall go out from (5.12). Evaluating the integrals (5.13) we obtain

$$\Gamma_{11} = 1,\ \Gamma_{1l} = -\frac{1}{4m^2} \text{ for } l > 1,$$
$$\Gamma_{i1} = -\frac{1}{4m^2} \text{ for } i > 1,\ \Gamma_{il} = 0 \text{ for } i, l > 1. \tag{5.15}$$

It follows from (5.12) that

$$\sum_{i=2}^{2M} \frac{a_i}{m^2} = \frac{4}{3} \text{ for } l = 1,$$
$$a_l = \frac{a_1}{4m} - \frac{1}{4m^2}(2k+1) \text{ for } l > 1. \tag{5.16}$$

From (5.16) we get

$$a_l \sum_{i=2}^{2M} \frac{1}{m^3} + \sum_{i=2}^{2M} \frac{2k+1}{m^4} = -\frac{16}{3}. \tag{5.17}$$

Since

$$\sum_{i=2}^{2M} \frac{1}{m^3} = 1 + (\frac{1}{8} + \frac{1}{8}) + (\frac{1}{64} + \frac{1}{64} + \frac{1}{64} + \frac{1}{64}) + \cdots + 2^J \frac{1}{2^{3J}}$$
$$= 1 + \frac{1}{2^2} + \frac{1}{2^4} + \cdots + \frac{1}{2^{2J}} = \frac{4}{3}(1 - \frac{1}{(2M)^2}) \tag{5.18}$$

and

$$\sum_{i=2}^{2M} \frac{2k+1}{m^4} = 1 + \sum_{k=0}^{1} \frac{2k+1}{2^4} + \sum_{k=0}^{2} \frac{2k+1}{2^8} + \cdots + \sum_{k=0}^{M-1} \frac{2k+1}{2^{4J}} = \frac{4}{3}(1 - \frac{1}{(2M)^2}) \tag{5.19}$$

it follows from (5.17) that

$$a_1 = \frac{5 + (2M)^{-2}}{1 - (2M)^{-2}} \tag{5.20}$$

The coefficients a_l for $l > 1$ can be calculated according to (5.16); the function $u(x)$ is evaluated from (5.7). Computations were carried out for different values of J. These results were compared with the exact solution

$$u_{ex} = x^2 - 5x - \frac{17}{6}. \tag{5.21}$$

The accuracy of the results were estimated by the local error function

$$e_J = \max_{1 \le l \le 2M} (| u(t_l) - u_{ex}(t_l) |), \tag{5.22}$$

where $t_l = (l - 0.5)/(2M)$. The errors e_J both for collocation and Galerkin method are given in Table 5.1.

Table 5.1 Error of solving (5.6) for $K = x + t,\ f(x) = x^2$

J	$2M$	e_J	
		Collocation	Galerkin
2	8	7.2E-2	1.1E-1
3	16	1.7E-2	2.6E-2
4	32	4.3E-3	6.5E-3
5	64	1.3E-3	1.6E-3

5.3 Eigenvalues and Eigenfunctions

Consider the first kind Fredholm equation

$$u(x) - \lambda \int_0^1 K(x,t)u(t)dt = 0, \tag{5.23}$$

where λ is a parameter. Equation (5.11) obtains now the form $a(H - \lambda G) = 0$. This system has a nontrivial solution only for some values of λ (eigenvalues). The corresponding functions $u(x)$ are eigenfunctions. The method of solution presented in Sect. 5.2 is also still applicable in this case. The details are explained by means of the following example.

Example 5.2: Here the equation

$$u(x) - \pi\lambda \int_0^1 \cos\pi(x+t)u(t)dt = 0 \tag{5.24}$$

is solved. Carrying out the integration (5.9) we obtain

$$G_1(x) = -\frac{2}{\pi}\sin\pi x \tag{5.25}$$

and

$$G_i(x) = \frac{1}{\pi}\left\{2\sin[\pi(x+\xi_2)] - \sin[\pi(x+\xi_1)] - \sin[\pi(x+\xi_3)]\right\}. \tag{5.26}$$

Here $\xi_1 = k/m,\ \xi_2 = (k+0.5)/m,\ \xi_3 = (k+1)/m,$. By applying trigonometric transformations, Eq. (5.26) can be rewritten in the form

$$G_i(x) = \frac{4}{\pi}\sin^2\frac{\pi}{4m}\sin[\pi(x+\frac{2k+1}{2m})],\ i = 2,3,\ldots,2M. \tag{5.27}$$

If we then want to apply the collocation method again, the matrix $G_{il} = G_i(x_l)$ is formed.

For solving the matrix equation $a(H - \lambda\pi G) = 0$ the MatLab program EIG was used. Now let us solve (5.24) with the Galerkin method. This procedure leads to the equation

$$\sum_{i=1}^{2M} a_i \int_0^1 h_i(x)h_l(x)dx = \pi\lambda \sum_{i=1}^{2M} a_i \Gamma_{il}, \tag{5.28}$$

which is view of (2.4) gets the form

$$\frac{a_l}{m} = \lambda\pi \sum_{i=1}^{2M} a_i \Gamma_{il},\ l = 1, \dots, 2M. \tag{5.29}$$

Taking into consideration (5.13) we get

$$\begin{aligned}
\Gamma_{11} &= -\frac{4}{\pi^2}, \\
\Gamma_{1l} &= \frac{8}{\pi^2}\sin^2\frac{\pi}{4m}\cos(\pi\frac{2k+1}{2m}),\ l = k+m+1 > 1, \\
\Gamma_{i1} &= \frac{8}{\pi^2}\sin^2\frac{\pi}{4m}\cos(\pi\frac{2k+1}{2m}),\ i = k+m+1 > 1, \\
\Gamma_{il} &= -\frac{16}{\pi^2}\sin^2\frac{\pi}{4m}\sin^2\frac{\pi}{4m}\cos\pi(\frac{2k+1}{2m}+\frac{2k+1}{2m}),\ i, l > 1.
\end{aligned} \tag{5.30}$$

For solving the system (5.29) again, the EIG program was used (this program also calculates the eigenfunctions u_1, u_2). The exact solution of the problem is

$$\lambda_1 = \frac{2}{\pi},\ \lambda_2 = -\frac{2}{\pi},\ u_1(x) = \cos\pi x,\ u_2(x) = \sin\pi x. \tag{5.31}$$

Some numerical results obtained by the Haar wavelet method are shown in Table 5.2. The eigenvalues $\lambda_1 = -\lambda_2$ (and their error in percents) are calculated both by the collocation and Galerkin methods. In the last column the error function (5.22) of the eigenfunctions is given (it turned out that the error was the same for both the collocation and Galerkin approach). Again high accuracy even with a small number of calculation points is stated.

It follows from Examples 5.1–5.2 that for the problems solved in Sects. 5.1–5.2 the collocation method is much simpler when compared with the Galerkin method. The accuracy of both methods is more or less the same.

Table 5.2 Eigenvalues of solving (5.24)and error of the normed eigenfunctions

J	$2M$	$\lambda_1 = -\lambda_2$		e_J
		Collocation	Galerkin	
2	8	0.6407 (0.64 %)	0.6449 (1.30 %)	1.9E-2
3	16	0.6376 (0.15 %)	0.6387 (0.32 %)	4.8E-3
4	32	0.6369 (0.04 %)	0.6371 (0.07 %)	1.2E-3
5	64	0.6367 (0.01 %)	0.6367 (0.01 %)	3.0E-4

5.4 Volterra Integral Equation

Consider the Volterra integral equation of the second kind

$$u(x) - \int_0^x K(x,t)u(t)dt = f(x), \ 0 \le x \le 1. \tag{5.32}$$

Its discrete form is

$$u(x_l) - \int_0^{x_l} K(x_l,t)u(t)dt = f(x_l), \tag{5.33}$$

where $x_l = (l-0.5)/(2M), \ l = 1, 2, \ldots, 2M)$ are the collocation points. We proceed as in Sect. 5.2, Eq. (5.10) holds. The matrix $G_{il} = G_i(x_l)$ is now defined as

$$G_{il} = \int_0^{t_l} K(x_l,t)h_i(t)dt. \tag{5.34}$$

By computing these integrals the following cases should be distinguished:

$$\begin{aligned}
&(i)\ G_{il} = 0, \text{ if } x_l < \xi_1,\\
&(ii)\ G_{il} = \int_{\xi_1}^{x_l} K(x_l,t)dt, \text{ if } \xi_1 \le x_l \le \xi_2,\\
&(iii)\ G_{il} = \int_{\xi_1}^{\xi_2} K(x_l,t)dt - \int_{\xi_2}^{x_l} K(x_l,t)dt, \text{ if } \xi_2 \le x_l \le \xi_3,\\
&(iv)\ G_{il} = \int_{\xi_1}^{\xi_2} K(x_l,t)dt - \int_{\xi_2}^{\xi_3} K(x_l,t)dt, \text{ if } \xi_3 \le x_l \le 1.
\end{aligned} \tag{5.35}$$

In view of (2.2) we have $\xi_1 = k/m, \ \xi_2 = (k+0.5)/m, \ \xi_3 = (k+1)/m$. In this and in the following sections the Galerkin solution is not considered.

Example 5.3: Let us take $K(x,t) = 3-x-t, \ f(x) = x^3$. Evaluating the integrals (5.35) we obtain

$$\begin{aligned}&(i)\ G_{il} = 0,\ \text{for } x_l < \xi_1,\\ &(ii)\ G_{il} = \tfrac{1}{2}(x_l - \xi_1)(6 - 3x_l - \xi_1) \text{ for } \xi_1 \le x_l \le \xi_2,\\ &(iii)\ G_{il} = (3 - x_l)(2\xi_2 - \xi_1 - x_l) + \tfrac{1}{2}(x_l^2 - 2\xi_2^2 + \xi_1^2) \text{ for } \xi_2 \le x_l \le \xi_3,\\ &(iv)\ G_{il} = \tfrac{1}{4m^2} \text{ for } \xi_3 \le x_l.\end{aligned} \tag{5.36}$$

To find the exact solution for the problem, (5.32) is differentiated twice with respect to x. Doing this we get the differential equation

$$u'' - (3 - 2x)u' + 3u = 6x \tag{5.37}$$

with the initial conditions $u(0) = u'(0) = 0$. This equation was integrated by the fourth order Runge-Kutta method and the result was taken for $u_{ex}(x)$. Computations gave the error the values:

$$e_2 = 1.2E - 2,\ e_3 = 3.1E - 3,\ e_4 = 7.9E - 4,\ e_5 = 2.0E - 4. \tag{5.38}$$

5.5 Integro-Differential Equation

Let us consider an equation of the following type

$$u'(x) + g(x)u(x) = \int_0^1 K(x,t)[\alpha u(t) + \beta u'(t)]dt + f(x), \tag{5.39}$$

where α, β are constants and $p(x)$, $f(x)$ are prescribed functions. To this equation belongs the initial condition $u(0) = \gamma$.

Developing the function $u'(t)$ into the Haar series we obtain

$$u'(t) = \sum_{i=1}^{2M} a_i h_i(t)\,,\ u(t) = \sum_{i=1}^{2M} a_i p_1(t) + u(0), \tag{5.40}$$

where $p_1(t) = \int_0^t h_i(t)dt$.

A substitution of (5.40) into (5.39) gives

$$\sum_{i=1}^{2M} a_i[h_i(x) + g(x)p_1(x) - \alpha R_i(x) - \beta G_i(x)] = -g(x)u(0) + \alpha u(0)Q(x) + f(x). \tag{5.41}$$

Here $G_i(x)$ is defined by (5.9), besides the following notations are used

$$Q(x) = \int_0^1 K(x,t)dt, \ R_i(x) = \int_0^1 K(x,t)p_1(t)dt. \tag{5.42}$$

Next we introduce the collocation points and pass to the matrices

$$H_{il} = h_i(x_l), \ G_{il} = G_i(x_l), \ Q_l = Q(x_l), \ R_{il} = R_i(x_l), \ V_{il} = g(x_l)p_1(x_l). \tag{5.43}$$

If g, f and Q are understood as $2M$-vectors, then the matrix form of (5.41) is

$$a(H + V - \alpha R - \beta G) = -\gamma g + \alpha\gamma Q + f. \tag{5.44}$$

Example 5.4: Consider the case $g(x) = 2x$, $K(x,t) = x + t$, $\alpha = 0$, $\beta = 1, \gamma = 0$, $f(x) = 2x^4 + 2x^3 + 3x^2 - 17/12$. Equation (5.44) gets the form

$$a(H + V - G) = f. \tag{5.45}$$

According to (5.9):

$$G_{il} = \begin{cases} x_l + 0.5 & \text{for } i = 1, \\ -1/(4m^2) & \text{for } i > 1. \end{cases} \tag{5.46}$$

After solving (5.45), the functions $u(x)$ and $u'(x)$ can be calculated from the formulas (5.40). The results of the computation were compared with the exact solution, which is $u_{ex} = x^2(1+x)$. The error estimate (5.22) was: $e_2 = 1.6E-2$, $e_3 = 4.0E-3$, $e_4 = 1.0E-3$, $e_5 = 2.6E-4$, $e_6 = 6.5E-5$.

5.6 Weakly Singular Integral Equations

The proposed method is applicable also for weakly singular equations

$$u(x) = \int_0^x \frac{K(x,t)u(t)}{(x-t)^\mu} dt + f(x), \ \mu \in (0.1), \ 0 \le t < x \le 1. \tag{5.47}$$

For illustration in the following the Abel integral equation is solved.

Example 5.5: Solve

$$\int_0^x \frac{u(t)}{\sqrt{x-t}} dt = 1, \ 0 \le t \le x \le 1. \tag{5.48}$$

By assuming

$$u(t) = \sum_{i=1}^{2M} a_i h_i(t), \tag{5.49}$$

and replacing it into (5.48) we get

$$\sum_{i=1}^{2M} a_i G_i(x_l) = 1, \; l = 1, \ldots, 2M \tag{5.50}$$

and

$$G_i(x) = \int_0^x \frac{h_i(t)}{\sqrt{x-t}} dt. \tag{5.51}$$

The matrix formulation of (5.50) is

$$aG = E \tag{5.52}$$

where $G_{il} = G_i(x_l)$ and E is a $2M$-dimensional unit vector. Evaluating the integrals $G_{il} = G_i(x_l)$ we obtain

$$\begin{aligned}
&(i)\; G_{il} = 0 \text{ for } x_l < \xi_1, \\
&(ii)\; G_{il} = 2\sqrt{x_l - \xi_1} \text{ for } \xi_1 \leq x_l \leq \xi_2, \\
&(iii)\; G_{il} = -4\sqrt{x_l - \xi_2} + 2\sqrt{x_l - \xi_1} \text{ for } \xi_2 \leq x_l \leq \xi_3, \\
&(iv)\; G_{il} = -4\sqrt{x_l - \xi_2} + 2\sqrt{x_l - \xi_1} + 2\sqrt{x_l - \xi_3} \text{ for } \xi_3 \leq x_l \leq 1.
\end{aligned} \tag{5.53}$$

The exact solution of (5.48) is

$$u_{ex} = \frac{1}{\pi\sqrt{x}}. \tag{5.54}$$

The computations which were carried out for the Haar wavelet approach showed that near the singularity $x = 0$ more collocation points must be taken. This situation is demonstrated in Fig. 5.1: only for $J = 7$ (or $2M = 256$) the accuracy of the wavelet solutions can be regarded as Smore or less satisfactory. If we move away from $x = 0$ the convergence is again good. So for the interval $0.1 \leq x \leq 1$ the error estimates (5.22) are $e_4 = 1.9E-2$, $e_5 = 1.3E-3$, $e_6 = 3.4E-4$, $e_7 = 8.8E-5$.

The convergence of the Haar series could be improved (and the number of collocation points reduced), if we make use of the segmentation method (Sect. 3.6) or apply the nonuniform Haar wavelets (Sect. 2.4).

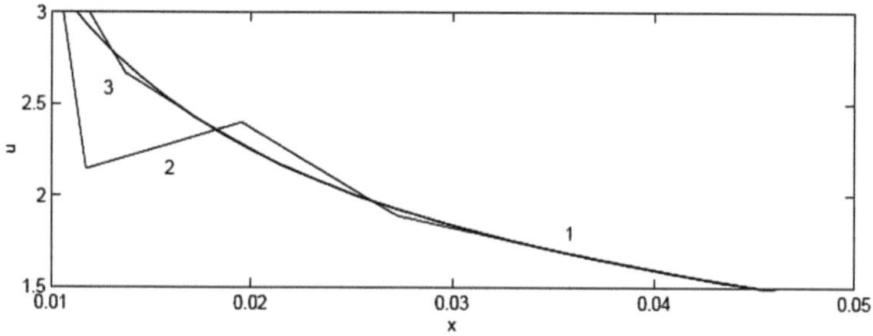

Fig. 5.1 Solution of the diffusion equation for $J = 3$

5.7 The Case of the Infinite Interval of Integration

Now we consider integral equations for which $B \to -\infty$ or $A \to \infty$. As a sample we discuss here the Volterra equation

$$u(x) = \int_0^\infty K(x,t)u(t)dt = f(x), \ \ (x,t) \in [0,\infty). \tag{5.55}$$

Since $u(t) \to 0$ for $x \to \infty$, we can choose a sufficiently large positive constant L and solve the problem for $(x,t) \in [0, L]$; here the method of nonuniform Haar wavelets with an increasing stepsize is recommended.

Example 5.6: Solve

$$u(x) = 4\int_0^\infty e^{-(x+t)}u(t)dt = (x-1)e^{-x}, \tag{5.56}$$

which has an exact solution

$$u_{ex}(x) = xe^{-x}. \tag{5.57}$$

The interval $[0, L]$ is divided into subintervals of length $\Delta x_i = \tilde{x}_i - \tilde{x}_{i-1}, i = 1, 2, \ldots, 2M$. It is assumed that $\Delta x_{i+1} = q\Delta x_i$, where $q > 1$ is a given constant. According to the formula (3.49) the grid points are

$$\tilde{x}_l = L\frac{q^l - 1}{q^{2M} - 1}, \ l = 1, 2, \ldots, 2M. \tag{5.58}$$

The collocation points x_l are calculated according to (2.9). Haar wavelet solution is sought in the form

Table 5.3 Error estimates for (5.56)

J	$2M$	$q = 1.0$		$q = 1.1$		$q = 1.2$	
		δ	σ	δ	σ	δ	σ
3	16	6.0E-2	4.4E-3	1.5E-2	1.3E-3	3.0E-3	3.4E-4
4	32	1.7E-2	8.0E-4	6.4E-4	5.1E-5	6.1E-4	7.0E-5
5	64	4.7E-3	1.4E-4	1.7E-4	6.4E-5	7.0E-4	7.3E-5

$$u(x) = \sum_{i=1}^{2M} a_i h_i(x). \tag{5.59}$$

Next the integral

$$G_i = \int_0^x e^{-t} h_i(t) dt \tag{5.60}$$

is calculated. In view of (2.1) we find

$$G_i = e^{-\xi_1(i)} + [1 + c(i)] e^{-\xi_2(i)} + C_3(i) e^{-\xi_3(i)}. \tag{5.61}$$

Making use of (5.59)–(5.60) and satisfying (5.56) in the collocation points x we obtain the matrix equation

$$aS = F, \tag{5.62}$$

where

$$S(i,l) = H(i,l) - 4G_i e^{-x(l)}, \quad F(l) = [x(l) - 1] e^{-x(l)}. \tag{5.63}$$

From here the vector a is calculated and the function $u(x)$ is found from (5.59). For estimating the accuracy of the obtained results, the error estimates (3.6)–(3.7) were used.

Computer simulations were carried out for $q = 1$ (uniform Haar solution) and $q = 1.1$, $q = 1.2$. For the parameter L the value $L = 10$ was taken. The results are presented in Table 5.3.

Calculations with $L = 20$ were also carried out but the results did not essentially differ from the data of Table 5.3. It follows from this table that results with variable stepsize are considerably more accurate as in the case of a constant stepsize.

5.8 Nonlinear Integro-Differential Equation

Consider Eq. (5.3), for simplicity sake, it is assumed that $A = 0,\ B = 1,\ u(0) = 0$. Satisfying it in the collocation points we obtain

$$\alpha u'(x_l) + \beta u(x_l) = \int_0^{x_l} K[x_l, t, u(t), u'(t)]dt + f(x_l), \tag{5.64}$$

where $l = 1, 2, \ldots, 2M$. The function $u'(t)$ is developed into the Haar series

$$u'(t) = \sum_{i=1}^{2M} a_i h_i(t). \tag{5.65}$$

By integrating (5.65) we obtain

$$u(t) = \sum_{i=1}^{2M} a_i p_i(t) + u(0). \tag{5.66}$$

If (5.65) and (5.66) are substituted into (5.64) and the integrations are carried out, a system of $2M$ nonlinear equations for calculating the wavelet coefficients a_i is obtained. This system is solved by the Newton method which brings us to the equation

$$\sum_{i=1}^{2M} [\alpha h_i(x_l) + \beta p_i(x_l) - \int_0^{x_l} \frac{\partial K}{\partial a_i} dt] \Delta a_i$$
$$= \alpha u'(x_l) - \beta u(x_l) + \int_0^{x_l} K dt + f(x_l), \; l = 1, 2, \ldots, 2M, \tag{5.67}$$

where

$$\frac{\partial K}{\partial a_i} = \frac{\partial K}{\partial u}\frac{\partial u}{\partial a_i} + \frac{\partial K}{\partial u'}\frac{\partial u'}{\partial a_i} = \frac{\partial K}{\partial u} p_i(t) + \frac{\partial K}{\partial u'} h_i(t). \tag{5.68}$$

The main problem is to evaluate the integrals in (5.67). This can be done in the following way. Let us denote

$$\varphi(l) = \int_0^{x_l} K dt, \; \psi(i,l) = \int_0^{x_l} \frac{\partial K}{\partial a_i} dt \tag{5.69}$$

and consider the subinterval $t \in [\tau_s, \tau_{s+1}]$, $s = 0, 1, 2, \ldots, 2M - 1$, where τ_s is the s-th grid point defined by $\tau_s = s/(2M)$. In each subinterval

$$\begin{aligned} h_i(t) &= h_i(t_s) = const., \\ p_i(t) &= p_i(\tau_s) + (t - \tau_s) h_i(t_s). \end{aligned} \tag{5.70}$$

Here t_s denotes the s-th collocation point

$$t_s = \tau_s + 0.5\Delta t. \tag{5.71}$$

Since $u'(t) = u'(t_s) = const$, we obtain

$$\begin{aligned} u(t) &= u(\tau_s) + (t - \tau_s)u'(t_s), \\ u(t_s) &= u(\tau_s) + 0.5u'(t_s)\Delta t. \end{aligned} \tag{5.72}$$

Next the following notations are introduced:

$$\begin{aligned}
\Delta G(x_l, t_s, u(t_s), u'(t_s)) &= \int_{\tau_s}^{\tau_{s+1}} K(x_l, t, u(t), u'(t_s))dt, \\
\Delta \tilde{G}(x_l, t_l, u(t_l), u'(t_l)) &= \int_{\tau_l}^{t_l} K(x_l, t, u(t), u'(t_l))dt, \\
\Delta G_u = \int_{\tau_s}^{\tau_{s+1}} \frac{\partial K}{\partial a_i} dt &= \int_{\tau_s}^{\tau_{s+1}} [\frac{\partial K}{\partial u} p_i(t) + \frac{\partial K}{\partial u'} h_i(t)]dt, \\
\Delta \tilde{G}_u = \int_{\tau_l}^{t_l} \frac{\partial K}{\partial a_i} dt &= \int_{\tau_l}^{t_l} [\frac{\partial K}{\partial u} p_i(t) + \frac{\partial K}{\partial u'} h_i(t)]dt.
\end{aligned} \tag{5.73}$$

Evaluating the integrals (5.69) for each subinterval $t \in [\tau_s, \tau_{s+1}]$ and summing up the results we obtain

$$\begin{aligned} \varphi(l) &= \sum_{s=1}^{l-1} \Delta G + \Delta \tilde{G}, \\ \psi(i, l) &= \sum_{s=1}^{l-1} \Delta G_u + \Delta \tilde{G}_u. \end{aligned} \tag{5.74}$$

It is convenient to put our results into the matrix form. For this purpose we introduce the row vectors $x = [x(l)]$, $a = (a_i)$, $\Delta a = (\Delta a_i)$, $u = [u(t_l)]$, $u' = [u'(t_l)]$, $f = [f(x_l)]$, $\varphi = [\varphi(l)]$ and $2M \times 2M$ matrices $H = [h_i(x_l)]$, $Q = [p_i(x_l)]$. Besides we denote

$$\begin{aligned} S &= \alpha H + \beta Q - \psi, \\ F &= -\alpha u' - \beta u + \varphi + f. \end{aligned} \tag{5.75}$$

The matrix form of the system (5.67) is now

$$\Delta a S = F, \tag{5.76}$$

which has the solution

$$\Delta a = F S^{-1}. \tag{5.77}$$

The following iteration process proceeds as shown in Sect. 3.5. We recommend taking the starting solution in the form $u'(0) = a_0,\ u(0) = a_0 x + u(0)$ and satisfy (5.64) in the single collocation point $x = 0.5$. This leads to the equation

$$\alpha a_0 + \beta(0.5a_0 + u(0)) = \int_0^{0.5} K[0.5, t, a_0 t + u(0), a_0]dt + f(0.5) \tag{5.78}$$

from which the coefficient a_0 can be evaluated. Estimates for the next step are $\hat{a}_1^{(1)} = a_0,\ \hat{a}_2^{(1)} = 0$ and

$$\begin{aligned}\hat{u}'^{(1)}(t) &= \hat{a}_1^{(1)} h_1(t) + \hat{a}_2^{(1)} h_2(t) = a_0,\\ \hat{u}^{(1)}(t) &= \hat{a}_1^{(1)} p_1(t) + \hat{a}_2^{(1)} p_2(t) + u(0) = a_0 t + u(0).\end{aligned} \tag{5.79}$$

These estimates are corrected by solving (5.77).

The Volterra integral equation

$$u(x) = \int_0^x K[x, t, u(t)]dt + f(x) \tag{5.80}$$

is a special case of (5.67). Our results remain applicable also for the Volterra equation if the following changes are carried out:

(i) $\alpha = 0,\ \beta = 1,\ \frac{\partial K}{\partial u'} = 0$
(ii) Equation (5.66) is replaced with $u = aH$.

Example 5.7: Consider the Volterra equation

$$u(x) = \int_0^x \frac{1 + u^2(t)}{1 + t^2} dt, \tag{5.81}$$

which has the exact solution $u_{ex} = x$. The wavelet solution is sought in the form

$$u(t) = \sum_{i=1}^{2M} a_i h_i(t). \tag{5.82}$$

Since $K = (1 + u^2)/(1 + t^2)$, it follows from (5.73) and (5.74) that:

Table 5.4 Error estimates δ, σ for problem (5.81)

J	$2M$	δ	σ
1	4	2.7E-2	–
2	8	1.6E-3	15.0
3	16	4.7E-4	3.7
4	32	1.3E-4	3.7
5	64	3.3E-5	3.9
6	128	8.4E-6	3.9

$$\varphi(l) = \sum_{s=1}^{l-1}[1 + u^2(t_s)A(s)] + 1 + u^2(t_l)a(l),$$
$$\psi(i, l) = 2\sum_{s=1}^{l-1} u(t_s)h_i(t_s)A(s) + 2[1 + u(t_l)h_i(t_l)]A(l), \; i, l = 1, 2, \ldots, 2M, \tag{5.83}$$

where

$$A(s) = \arctan \tau_{s+1} - \arctan \tau_s,$$
$$A(l) = \arctan t_l - \arctan \tau_l. \tag{5.84}$$

We start our solution with $u = a_0 = const$. Satisfying (5.78) we get the equation

$$a_0^2 - \frac{a_0}{\arctan 0.5} + 1 = 0, \tag{5.85}$$

which has two roots $a_0^{(1)} = 0.6747$ and $a_0^{(2)} = 1.4821$. It follows from the calculations that the value $a_0^{(2)}$ brings to a non-convergent iteration process and therefore we take $a_0 = 0.6747$. Next two collocation points $x_1 = 0.25$ and $x_2 = 0.75$ are taken and the estimates for a and u are $\hat{a} = (a_0, 0),\ \hat{u} = (a_0, a_0)$. Correcting these values with the aid of (5.77) we obtain $a^{(1)} = (0.523,\ 0.323)$ and $u^{(1)} = (0.199,\ 0.846)$; the error function (5.22) is $e^{(1)} = 0.096$. Results of the following iterations are shown in Table 5.4. In Table 5.4 the quantity $\sigma = \delta_{J-1}/\delta_J$, which characterizes the rate of convergence, is introduced.

Example 5.8: Consider the integro-differential equation

$$u'(x) = 1 + \int_0^x u(t)u'(t)dt, \; 0 \le x \le 1, \; u(0) = 0. \tag{5.86}$$

The exact solution of (5.86) is

$$u_{ex} = \sqrt{2} \tan \frac{x}{\sqrt{2}}. \tag{5.87}$$

Table 5.5 Error estimates δ, σ for (5.86)

J	$2M$	δ	σ
1	4	4.2E-3	–
2	8	3.3E-3	1.3
3	16	1.1E-3	3.0
4	32	3.0E-4	3.7
5	64	8.9E-5	3.3
6	128	2.6E-5	3.4

We shall seek the wavelet solution in the form (5.65) and (5.66); according to (5.73) and (5.74) we find

$$\varphi(l) = \Delta t \sum_{s=1}^{l-1} u(t_s)u'(t_s) + \frac{\Delta t}{2} u(t_l)u'(t_l),$$
$$\psi(i,l) = \Delta t \sum_{s=1}^{l-1} [u(t_s)h_i(t_s) + u'(t_s)p_i(t_s)] + \frac{\Delta t}{2}[u(t_l)h_i(t_l) + u'(t_l)p_i(t_l)]. \tag{5.88}$$

Equations (5.75) and (5.76) obtain the form $\Delta a S = F$, where $S = H - \psi$, $F = -u' + \varphi + E$ (symbol E denotes the $2M$ dimensional unit vector).To start with we take $u' = a_0$, $u = a_0 x$. Satisfying (5.86) for $x = 0.5$ the equation $a_0^2 - 8a_0 + 8 = 0$ is obtained. It has two roots from which we shall take $a_0 = 1.172$ (the other root leads to a non-converging process).Correcting the estimates $\hat{a} = (a_0, 0)$, $\hat{u}' = (a_0, a_0)$, $\hat{u} = 0.25(a_0, 3a_0)$ with the aid of (5.77) we find $a = (1.237, -0.190)$, $u' = (1.047, 1.426)$, $u = (0.262, 0.880)$ the error of this approximation is $e = 0.051$. Results for the next approximations are presented in Table 5.5.

This example was also solved with the aid of the Adomian's decomposition method by El-Sayed and Ab del-Aziz [10]. The authors denote that their solution is more simple and easy to use. However, we would like to note that the accuracy of the results may be insufficient. So it follows from Table 5.3 of the paper by El-Sayed and Abdel-Aziz that $u = 1.0973$ for $x = 1$ while the exact value is $u = 1.2085$; consequently, the error is 9.2 %.

5.9 Application of the Integro-Differential Equations for Solving Boundary Value Problems of ODE

The method of solution proposed in Sect. 5.8 can be applied for solving boundary value problems of ordinary differential equations. To illustrate this, let us solve the two-point boundary value problem for the differential equation

$$u'' = K(x, u, u') \tag{5.89}$$

with the boundary conditions $u(0) = u_0,\ u(1) = u_1$. By integrating (5.89) we get the integro-differential equation

$$u'(x) = \int_0^x K[t, u(t), u'(t)]dt + u'(0),\ 0 \le x \le 1. \tag{5.90}$$

Again we seek the solution in the form (5.65) and (5.66). Since $p_i(1) = 1$ for $i = 1$ and $p_i(1) = 0$ for $i \neq 1$ it follows from (5.66) that $a_1 = u_1 - u_0$, consequently, the first wavelet coefficient is fixed.According to (5.65)

$$u'(0) = \sum_{i=1}^{2M} a_i h_i(0) \tag{5.91}$$

where

$$h_i(0) = \begin{cases} 1 \text{ if } i = 1, \\ 1 \text{ if } i = 2^J + 1, \quad J = 0, 1, 2, \ldots \\ 0 \text{ elsewhere}. \end{cases} \tag{5.92}$$

Due to the fact that a_1 is fixed and $\Delta a_1 = 0$, some changes in the system (5.76) must be executed. Let the symbol $\tilde{a}$ denote the vector a for which the first component is deleted. In a similar way, the $\hat{S}$ is a reduced matrix in which the first row and column are deleted. Instead of (5.75)–(5.76) we get now

$$\Delta\tilde{a}\tilde{S} = \tilde{F}, \tag{5.93}$$

where

$$\begin{aligned} \tilde{S} &= \tilde{H} - \tilde{\psi} - \tilde{E} \otimes \tilde{h}(0)^T, \\ \tilde{F} &= \tilde{u}' + \tilde{u}'(0) + \tilde{\varphi} + \tilde{f}. \end{aligned} \tag{5.94}$$

Here $\tilde{E}$ is a $2M - 1$ dimensional unit vector, and $\otimes$- denotes the Kronecker tensor product. The following solution proceeds according to the algorithm presented in Sect. 5.8. We generate the vector $a = (B - A,\ a_2, \ldots, a_{2M})$ and evaluate $u(x),\ u'(x)$ for the next approximation from (5.65) and (5.66).

Example 5.9: Consider the boundary value problem

$$u'' - uu' = \frac{1}{4\sqrt{(1+x)^3}} - \frac{1}{2},\ u(0) = 1,\ u(1) = \sqrt{2}, \tag{5.95}$$

which has the exact solution $u = \sqrt{1+x}$. By integrating (5.95) we obtain

$$u' = \int_0^x uu'dt + u'(0) - \frac{1}{2}(1+x) + \frac{1}{2\sqrt{1+x}}. \tag{5.96}$$

Table 5.6 Error estimates δ, σ for (5.95)

J	$2M$	δ	σ
1	4	3.2E-3	–
2	8	1.1E-3	2.9
3	16	3.2E-4	3.5
4	32	8.6E-5	3.7
5	64	2.2E-5	3.9
6	128	5.6E-6	3.9

It follows from boundary conditions that $a_1 = u(1) - u(0) = \sqrt{2} - 1$. We start our solution with evaluating the quantities $\varphi(l)$ and $\psi(i, l)$. Making use of (5.73) and (5.74) we obtain

$$\begin{aligned}\varphi(l) &= \Delta t \sum_{s=1}^{l-1} u(t_s)u'(t_s) + \frac{\Delta t}{2} u'(t_l)[u(t_l) - \frac{\Delta t}{4} u'(t_l)],\\ \psi(i, l) &= \Delta t \sum_{s=1}^{l-1} [u(t_s)h_i(t_s) + u'(t_s)p_i(t_s)]\\ &+ \tfrac{\Delta t}{2}[u(t_l)h_i(t_l) + u'(t_l)p_i(t_l)] - (\tfrac{\Delta t}{2})^2 u'(t_l)h_i(t_l).\end{aligned} \tag{5.97}$$

Since now a_1 is fixed we take the starting solution in the form

$$\begin{aligned}u'(t) &= a_1 + a_2 h_2(t),\\ u(t) &= a_1 t + a_2 p_2(t) + u(0).\end{aligned} \tag{5.98}$$

Replacing (5.98) into (5.96) and satisfying this equation in the point $x = 0.5$ we get the quadratic equation

$$a_2^2 + 2(5 + \sqrt{2})a_2 + 2\sqrt{2} + \frac{8}{\sqrt{6}} - 7 = 0. \tag{5.99}$$

This equation has two roots from which fits to us the root $a_2 = 0.0702$. The estimates for the first approximation are $\hat{a} = (a_1,\ a_2) = (0.4142,\ 0.0702,\ \hat{u}' = (0.4943,\ 0.3539),\ \hat{u} = (1.124,\ 1.336)$. Correcting the values with the aid of (5.93) and (5.97) we find $a^{(1)} = (0.424,\ 0.056)$, $u^{(1)} = (1.120,\ 1.332)$, $u'^{(1)} = (0.480,\ 0.368)$ with the error $e^{(0)} = 0.009$. Results of the subsequent approximations are presented in Table 5.6.

5.10 Nonlinear Fredholm Integral Equation

Here we consider the equation

$$u(x) = \int_0^1 K[x, t, u(t)]dt + f(x), \quad (x, t) \in [0, 1]. \tag{5.100}$$

The method of solution presented in Sect. 5.8 remains valid if we carry out the following changes:

(i) $\alpha = 0,\ \beta = 1,\ \frac{\partial K}{\partial u'} = 0$
(ii) Equations (5.65)–(5.66) are replaced by (5.49)
(iii) The upper bound of the integrals (5.64), (5.67), (5.69) is changed to 1.

Now we have

$$\begin{aligned}\varphi(l) &= \sum_{s=1}^{2M}[G(x_l, \tau_{s+1}, u(t_s)) - G(x_l, \tau_s, u(t_s))],\\ \chi(i, l) &= \sum_{s=1}^{2M}[G_u(x_l, \tau_{s+1}, u(t_s)) - G_u(x_l, \tau_s, u(t_s))]h_i(t_s).\end{aligned} \tag{5.101}$$

Here the notations

$$\begin{aligned}G(x_l, t, u(t_s)) &= \int K(x_l, t, u(t_s))dt,\\ G_u(x_l, t, u(t_s)) &= \int \frac{\partial K(x_l, t, u(t_s))}{\partial u}dt\end{aligned} \tag{5.102}$$

were used Eq. (5.75) obtain the form

$$S = H - \psi,\ F = -u + \varphi + f. \tag{5.103}$$

For further details and complementary examples consult [20].

Example 5.10: Solve the equation

$$u(x) = x\int_0^1 t\sqrt{u(t)}dt + 2 - \frac{x}{3}(2\sqrt{2} - 1) - x^2, \tag{5.104}$$

which has the exact solution $u_{ex} = 2 - x^2$.

Evaluating the functions φ and χ, we find according to (5.101):

$$\begin{aligned}\varphi(l) &= x_l \Delta t \sum_{s=1}^{2M} t_s \sqrt{u^{(\nu)}(t_s)}, \\ \chi(i,l) &= 0.5 x_l \Delta t \sum_{s=1}^{2M} \frac{t_s h_i(t_s)}{\sqrt{u^{(\nu)}(t_s)}}.\end{aligned} \tag{5.105}$$

Here we get two solutions $a^{(0)} = 0.756$, $a_*^{(0)} = 1.253$. Let us consider the first solution. The approximation for $M = 1$ gives $a^{(1)} = (1.695, 0.246)$ and $u^{(1)} = (1.941, 1.448)$ with the error $\delta = 0.01$. Error estimates for the following levels of resolution J are $\delta_2 = 2.7E-3$, $\delta_3 = 1.1E-3$, $\delta_4 = 3.7E-4$, $\delta_5 = 1.1E-4$, $\delta_6 = 3.1E-5$. Practically the same results are achieved if we start with the value $a_*^{(0)} = 1.253$. Correcting it with the aid of (5.76) we again get $u_*^{(1)} = (1.941, 1.448)$ and, of course, the following approximations also coincide.

5.11 Related Papers

Beginning from 1991, the wavelet methods have been applied for solving integral equations. A short survey about the earlier papers can be found in [19]. Evidently the first papers in which the Haar wavelet method was applied belong to Maleknejad et al. [23–25] and to Lepik and Tamme [19]. In these papers linear integral equations are discussed. If the papers by Maleknejad et al. are concerned with the solution of Fredholm equations, then the scope of the paper [19] is wider—here Fredholm and Volterra equations, integro-differential equations and weakly singular equations are solved.

In the subsequent papers, Maleknejad and his co-workers applied the haar method for solving Fredholm and Volterra integral equations [21, 31, 34, 35]. Systems of linear integro-differential equations were considered in [24]. Different integral equations were solved by Haar wavelet method in [15–20]. Numerical solution of integral equations was presented by Shahsavaran [37].

In some papers the Haar wavelet method was used for solving nonlinear integral equations. Nonlinear Fredholm equations were discussed by Maleknejad and Karami [21] and by Babolian and Shahsavaran [6]. Ordokhani [33] solved nonlinear Volterra-Fredholm-Hammerstein equations.

Aziz and Siraj-ul-Islam [3] applied a solution of the nonlinear Fredholm and Volterra integral equations on a two-dimensional Haar basis. Babolian et al. [4] developed a new method for solution of nonlinear two-dimensional integral equations.

Now let us pass to the other wavelet families.

Walsh functions for solving integral equations were used by Hsiao and Chen [13]. Sepehrian and Razzaghi [36] applied the single-term Walsh series for solving Volterra integro-differential equations. Sloss and Blyth [39] used the Walsh function method for solving nonlinear Volterra integral equations.

Somewhat surprising is the great number of papers based on the block-pulse functions (BPF). Maleknejad with co-authors made use of the BPF for solving Volterra and Fredholm integral equations [27, 28]. The Nonlinear Fredholm integral equation was discussed in [37]. Two-dimensional BPF were applied for solving double integral equations by Maleknejad et al. [26, 30], Maleknejad and Rahimi [27], Mirzaee and Hadadiyan [32], Babolian et al. [5]. In these papers operational matrix of integration for 2D-BPF was composed and the problem is reduced for solving a system of nonlinear algebraic equations.

Periodized Daubechies wavelets based on the Galerkin method for solving different integral equations were used in [41]. Cattani [8] applied the Shannon wavelets for solving integro-differential equations. Harmonic wavelets were applied as basis functions in solution of Fredholm integral equations of the second kind by Cattani and Kudreyko [9].

In several papers the Legendre's wavelets are used. Khellat and Yousefi [14] and also Yousefi and Razzaghi [43] put together the Legendre mother wavelet operational matrix of integration. Fredholm integral equations were solved by Maleknejad and Sohrabi [29] and also Shang and Han [38]. A Legendre wavelets direct method for solving linear integro-differential equations was proposed by Abbas et al. [1]. Yousefi [42] applied Legendre wavelets for solution of Abel's integral equation. Nonlinear integral equations were analysed by Yousefi and Razzaghi [43]. Hybrid functions, which consist of block-pulse functions plus Legendre polynomials, were proposed for solution of integral equations by Hsiao [12]. In some papers the solution of the integral equations is based on the Chebyshev wavelets [5, 7, 11, 40]. Solution by using Coifman wavelets are presented in [22].

References

1. Abbas, Z., Vahdati, S., Atan, K., Long, N.: Legendre multi-wavelets direct method for linear integro-differential equations. Appl. Math. Sci. **3**, 693–700 (2006)
2. Atkinson, K.E.: The Numerical Solution of Integral Equations of the Second Kind. Cambridge University Press, Cambridge (1997)
3. Aziz, I.: Siraj-ul-Islam: New algorithms for the numerical solution of nonlinear Fredholm and Volterra integral equations using Haar wavelets. J. Comput. Appl. Math. **239**, 333–345 (2013)
4. Babolian, E., Bazm, S., Lima, P.: Numerical solution of nonlinear two-dimensional integral equations using rationalized Haar functions. Commun. Nonlinear Sci. Numer. Simul. **16**, 1164–1175 (2011)
5. Babolian, E., Maleknejad, K., Mordad, M., Rahimi, B.: A numerical method for solving FredholmVolterra integral equations in two-dimensional spaces using Block Pulse functions and an operational matrix. J. Comput. Appl. Math. **235**, 3965–3971 (2011)
6. Babolian, E., Shahsavaran, A.: Numerical solution of nonlinear Fredholm integral equations of the second kind using Haar wavelets. J. Comput. Appl. Math. **225**, 87–95 (2009)
7. Biazar, J., Ebrahimi, H.: Chebyshev wavelets approach for nonlinear systems of Volterra integral equations. Comput. Math. Appl. **63**, 608–616 (2012)
8. Cattani, C.: Shannon wavelets for the solution of integro-differential equations. Mathematical Problems in Engineering **2010** (2010). doi:10.1155/2010/408418
9. Cattani, C., Kudreyko, A.: Harmonic wavelet method towards solution of the Fredholm type integral equations of the second kind. Appl. Math. Comput. **215**, 4164–4171 (2010)

10. El-Sayed, S., Abdel-Aziz, M.: A comparison of Adomian's decomposition method and wavelet-Galerkin method for solving integro-differential equations. Appl. Math. Comput. **136**, 151–159 (2003)
11. Ghasemi, M., Kajani, M.: Numerical solution of time-varying delay systems by Chebyshev wavelets. Appl. Math. Model. **35**, 5235–5244 (2011)
12. Hsiao, C.: Hybrid function method for solving Fredholm and Volterra integral equations of the second kind. Comp. Appl. Math. **230**, 59–68 (2009)
13. Hsiao, C., Chen, C.: Solving integral equations via Walsh functions. Comput. Electr. Eng. **6**, 279–292 (1979)
14. Khellat, F., Yousefi, S.: The linear Legendre mother wavelets operational matrix of integration and its application. J. Franklin Inst. **343**, 181–190 (2006)
15. Lepik, Ü.: Haar wavelet method for nonlinear integro-differential equations. Appl. Math. Comp. **176**, 324–333 (2006)
16. Lepik, Ü.: Application of the Haar wavelet transform to solving integral and differential equations. Proc. Estonian Acad. Sci. Phys. Math. **56**, 28–46 (2007)
17. Lepik, Ü.: Solving differential and integral equations by the Haar wavelet method; revisited. Int. J. Math. Comput. **1**, 43–52 (2008)
18. Lepik, Ü.: Solving integral and differential equations by the aid of nonuniform Haar wavelets. Appl. Math. Comp. **198**, 326–332 (2008)
19. Lepik, Ü., Tamme, E.: Application of Haar wavelets for solution of linear integral equations. In: H. Akca, A. Boucherif, V. Covachev (eds.) Dynamical Systems and Applications, pp. 494–507 (2005)
20. Lepik, Ü., Tamme, E.: Solution of nonlinear Fredholm integral equations via the Haar wavelet method. Proc. Estonian Acad. Sci. Phys. Math. **56**, 17–27 (2007)
21. Maleknejad, K., Karami, M.: Numerical solution of non-linear Fredholm integral equations by using multiwavelets in the Petrov-Galerkin method. Appl. Math. Comput. **168**, 102–110 (2005)
22. Maleknejad, K., Lotfi, T., Rostami, Y.: Numerical computational method in solving Fredholm integral equations of the second kind by using Coifman wavelet. Appl. Math. Comput. **186**, 212–218 (2007)
23. Maleknejad, K., Mirzaee, B.: Using rationalized Haar wavelet for solving linear integral equations. Appl. Math. Comp. **160**, 579–589 (2005)
24. Maleknejad, K., Mirzaee, F.: Numerical solution of linear Fredholm integral equations system by rationalized Haar functions method. Int. J. Comput. Math. **80**, 1397–1405 (2003)
25. Maleknejad, K., Mirzaee, F., Abbasbandy, S.: Solving linear integro-differential equations system by using rationalized Haar functions method. Appl. Math. Comput. **155**, 317–328 (2004)
26. Maleknejad, K., Mollapourasl, R., Alizadeh, M.: Numerical solution of Volterra type integral equation of the first kind with wavelet basis. Appl. Math. Comput. **194**, 400–405 (2007)
27. Maleknejad, K., Rahimi, B.: Modification of Block Pulse functions and their application to solve numerically Volterra integral equation of the first kind. Commun. Nonlinear Sci. Numer. Simul. **16**, 2469–2477 (2011)
28. Maleknejad, K., Shahrezaee, M., Khatami, H.: Numerical solution of integral equations system of the second kind by BlockPulse functions. Appl. Math. Comput. **166**, 15–24 (2005)
29. Maleknejad, K., Sohrabi, S.: Numerical solution of Fredholm integral equations of the first kind by using Legendre wavelets. Appl. Math. Comput. **186**, 836–843 (2007)
30. Maleknejad, K., Sohrabi, S., Baranji, B.: Application of 2D-BPFs to nonlinear integral equations. Commun. Nonlinear Sci. Numer. Simul. **15**, 527–535 (2010)
31. Mirzaee, F.: Numerical computational solution of the linear Volterra integral equations system via rationalized Haar functions. J. King Saud Univ. Sci. **22**, 265–268 (2010)
32. Mirzaee, F., Hadadiyan, E.: Approximate solutions for mixed nonlinear Volterra-Fredholm type integral equations via modified Block-Pulse functions. J. Assoc. Arab Univ. Basic Appl. Sci. **12**, 65–73 (2012)

33. Ordokhani, Y.: Solution of nonlinear Volterra-Fredholm-Hammerstein integral equations via rationalized Haar functions. Appl. Math. Comput. **180**, 436–443 (2006)
34. Rabbani, M., Maleknejad, K., Aghazadeh, N., Mollapourasl, R.: Computational projection methods for solving Fredholm integral equation. Appl. Math. Comput. **191**, 140–143 (2007)
35. Reihani, M., Abadi, Z.: Rationalized Haar functions method for solving Fredholm and Volterra integral equations. J. Comput. Appl. Math. **200**, 12–20 (2007)
36. Sepehrian, B., Razzaghi, M.: Single-term Walsh series method for the Volterra integro-differential equations. Eng. Anal. Bound. Elem. **28**, 1315–1319 (2004)
37. Shahsavaran, A., Shahsavaran, A.: Properties of BPFs for approximating the solution of non-linear Fredholm integro differential equation. Appl. Math. Sci. **6**, 1563–1569 (2012)
38. Shang, X., Han, D.: Numerical solution of Fredholm integral equations of the first kind by using linear Legendre multi-wavelets. Appl. Math. Comput. **191**, 440–444 (2007)
39. Sloss, B., Blyth, W.: A Walsh function method for a non-linear Volterra integral equation. J. Franklin Inst. **340**, 25–41 (2003)
40. Sohrabi, S.: Comparison Chebyshev wavelets method with BPFs method for solving Abel's integral equation. Ain Shams Eng. J. **2**, 249–254 (2011)
41. Xiao, J., Wen, L., Zhang, D.: Solving second kind Fredholm integral equation by periodic wavelet Galerkin method. Appl. Math. Comput. **175**, 508–518 (2006)
42. Yousefi, S.: Numerical solution of Abel's integral equation by using Legendre wavelets. Appl. Math. Comput. **175**, 574–580 (2006)
43. Yousefi, S., Razzaghi, M.: Legendre wavelets method for the nonlinear Volterra-Fredholm integral equations. Math. Comput. Simul. **70**, 1–8 (2005)

Chapter 6
Evolution Equations

6.1 Problem Statement and Methods of Solution

Now let us pass to partial differential equations (PDEs). In this chapter we deal with *evolution equations* of the following type

$$\alpha\ddot{u}(t,x) + \beta\dot{u}(t,x) = F[t, x, u(t,x), u'(t,x), u''(t,x)]. \tag{6.1}$$

Here α, β are given constants, t and x are arguments; usually t denotes time and x is a spacial coordinate, F is a given function, $u(t,x)$ is the function being sought. Dots and primes denote differentiation with regard to the arguments t and x, respectively.

To (6.1) belong the initial and boundary conditions. We confine us to the case where $0 \le t \le T$ and $0 \le x \le L$. The initial conditions are $u(0,x) = u_0(x)$; if $\alpha \ne 0$ in addition $\dot{u}(0,x) = \dot{u}_0(x)$ must be satisfied. The boundary conditions we take in the form $u(t,0) = f(t),\ u(t,L) = g(t)$.

The evolution equations are essential for solving many problems in mathematical physics, computational physics, mechanics, engineering and other disciplines. Different methods of solution have been worked out.

In this chapter, solutions based on the Haar wavelets are presented. Here three main approaches are possible.

(i) Making use of the two-dimensional Haar wavelets; this method is considered in the next chapter.

(ii) Chen and Hsiao [8] suggested seeking the solution of (6.1) in the case $\alpha = 0$ in the matrix form

$$\dot{u}(x,t) = a(x)H(t), \tag{6.2}$$

where $a(x)$ is $2M$-dimensional vector and $H(t)$ the Haar matrix. Equation (6.2) is integrated with respect to t and differentiated twice with respect to x. The obtained values of u, u', u'' are substituted into (6.1). By doing this we get an ODE for

Ü. Lepik and H. Hein, *Haar Wavelets*, Mathematical Engineering,
DOI: 10.1007/978-3-319-04295-4_6, © Springer International Publishing Switzerland 2014

calculating the wavelet coefficients $a(x)$. If the function F is linear, this equation can be solved analytically.

This approach is applied for solving the diffusion equation in Sect. 6.2.

(iii) Let us divide the time interval into N parts of equal length Δt and develop the highest derivative $\ddot{u}''$ in (6.1) into Haar series. This term has the form

$$\ddot{u}'' = \sum_{i=1}^{2M} a_i(t_s) h_i(x), x \in [0, L], \tag{6.3}$$

where $t_s = s\Delta t, \ s = 1, 2, \ldots, N$.

We assume that $a_s(t) = const$ in the interval $t \in [t_s, t_{s+1}]$. This enables us to integrate (6.1) as a function of time. Integration with respect to the space variable x is accomplished by the Haar wavelet method. Repeating this procedure for each subinterval we can evaluate the wavelet coefficients as time functions $a_i(t), \ i = 1, 2, \ldots, 2M$ for the whole interval $t \in [0, T]$. This method is applied for the solution of two problems in Sects. 6.3–6.4.

Results of this Chapter were published in the papers [25–27].

6.2 Diffusion Equation

The one-dimensional diffusion equation in the unit interval has the form

$$\dot{u} = Au'', \ 0 \le x \le 1, \ 0 \le t \le 1. \tag{6.4}$$

To this equation belong boundary conditions $u(0, t) = u(1, t) = 0$ and the initial condition $u(x, 0) = u_0(x)$. The same equation describes propagation of heat in a rod and therefore is also called the heat equation. Equation (6.4) is the simplest of the evolution equations and has been a subject of research for more than 150 years. Different methods of analytical and numerical solution have been proposed.

In the following we present a solution based on the Haar wavelets. Following the treatment of Chen and Hsiao (1997) [8] we seek the solution in the matrix form (6.2).

This equation is integrated in regard to t and differentiated twice with respect to the space coordinate x; by doing this we find

$$u(x, t) = a(x)P_1(t) + u(x, 0)E, \tag{6.5}$$

$$u''(x, t) = a''(x)P_1(t) + u''(x, 0)E, \tag{6.6}$$

where E is a $2M$ dimensional unit vector.

Substitution into (6.4) gives

$$aH = A[a''P_1 + u''(x, 0)E]. \tag{6.7}$$

This is for x an ODE and can be solved in a traditional way.

Let us specify $u_0 = x(1-x)$. The Eq. (6.7) can be put into the form

$$a'' - \omega^2 a = S, \tag{6.8}$$

where

$$\omega^2 = \frac{1}{A} H / P_1 \tag{6.9}$$

and

$$S = 2E/P_1 = 4[0, 0, \ldots, 0, 2M, \ldots, 2M]. \tag{6.10}$$

Solution of the matrix Eq. (6.8) is

$$a(x) = \gamma expm(\omega x) + \delta expm(-\omega x) - 2AY, \tag{6.11}$$

where $Y = [1, 0, \ldots, 0]$, besides $\omega = sqrtm(\omega^2)$. Here the symbols *expm* and *sqrtm* denote exponential and square root matrices, respectively. The integration constants are calculated from the boundary conditions $u(0,t) = u(1,t) = 0$. In view of (6.5) we find $a(0) = a(1) = 0$ and

$$\gamma = -2AY \frac{expm(-\omega)}{\Delta}, \ \delta = 2A \frac{expm(\omega)}{\Delta}, \tag{6.12}$$

where $\Delta = expm(\omega) - expm(-\omega)$.

The function $u(x,t)$ is evaluated from (6.5).

With the purpose of estimating the accuracy of the obtained results, computer simulation was carried out for $A = 0.2, \ x = 0.5, \ t = 0.25$. The results were compared with the Fourier solution

$$u(x,t) = \sum_n d_n exp(-\pi^2 n^2 At) \sin n\pi x, \tag{6.13}$$

where $d_n = (2/n\pi)^3, \ n = 1, 3, 5, \ldots$. For $J = 5$ (16 collocation points) we found $u(0.5, 0.25) = 0.1676$, while the Fourier solution gives the value 0.1574. Consequently, the error of the wavelet solution is 6.5 %. Attempts to raise the accuracy of solution by increasing J were unsuccessful, since for $J \geq 4$ the solution became unstable. The same situation occurred in the case of other values of A.

The reason for the instability lies in the fact that to get the solution we must invert the matrix P_1 (see (6.9), (6.10)). By increasing J the matrix P_1 turns nearly singular and the inversion of it brings great errors. To illustrate it, we calculated the determinant of this matrix and found that $\mid P_1 \mid = 2.7E - 20$ for $J = 3$, $\mid P_1 \mid = 3.4E - 49$ for $J = 4$ and $\mid P_1 \mid = 6.6E - 81$ for $J = 5$.

From this we can conclude that for getting high accuracy of the Haar wavelet results, we must use such algorithms in which inversion of the integral matrices

$P_1, P_2, \ldots$ is avoided. This can be done in different ways, e.g. we could apply the segmentation method or the piecewise constant approximation method, which were discussed in Sect. 3.6. Surprisingly good results were given by the PCA (also called single-term Haar wavelet method): for $\Delta t = 0.002$ and $N_t = 32$ the PCA gave $u_{max} = 0.1575$ (Fourier solution gives 0.1524) for $A = 2$ (for details consult [25], Sect. 6). An advantage of the PCA method is that it remains stable if the number of calculation points is increased.

6.3 Burgers Equation

Consider the Burgers equation

$$\frac{\partial u}{\partial t_*} + u\frac{\partial u}{\partial x_*} = \nu_* \frac{\partial^2 u}{\partial x_*^2}, \quad x^* \in [-1, 1], \quad t_* > 0 \tag{6.14}$$

with the boundary and initial conditions $u(\pm 1, t) = 0$, $u(x, 0) = \sin \pi x_*$. The constant ν_* is the viscosity coefficient. Burgers equation has proved to be a touchstone for new numerical methods of solution. This is caused by two reasons. First Burgers equation is the simplest nonlinear PDE incorporating both diffusion and advection. The second reason is that for a small viscosity ν^*, the solution develops into a saw-tooth wave at the origin. Performance of a numerical method can be judged from its ability to resolve the large gradient region that develops in the solution.

An overview of the conventional methods and obtained numerical results can be obtained from some papers about the Burgers equation (see e.g. Basdevant et al. [3], Jiwari [23]). There exists also an analytical solution, but it is quite difficult to compute.

It is convenient to put (6.14) into the normalized form, this can be done by changing the variables $x_* = 2x - 1$, $t_* = t$, $\gamma_* = 4\gamma$. Now we get the problem

$$\begin{aligned} \dot{u} + \frac{1}{2}uu' = \gamma u'', \ x \in [0, 1], \ t > 0, \\ u(x, 0) = \sin 2\pi x, \ u(0, t) = u(1, t) = 0. \end{aligned} \tag{6.15}$$

The Haar wavelet solution is sought in the form

$$\dot{u}''(x, t) = \sum_{i=1}^{2M} a_s(i) h_i(x), \tag{6.16}$$

where the row vector a_s is constant in the subinterval $t \in [t_s, t_{s+1}]$.

By integrating (6.16) with respect to t in the limits $[t_s, t]$ and twice with respect to x in the limits $[0, x]$ we obtain

$$
\begin{aligned}
u''(x,t) &= (t-t_s)\sum_{i=1}^{2M} a_s(i)h_i(x) + u''(x,t_s)\\
u'(x,t) &= (t-t_s)\sum_{i=1}^{2M} a_s(i)p_{1,i}(x) + u'(x,t_s) - u'(0,t_s) + u'(0,t)\\
u(x,t) &= (t-t_s)\sum_{i=1}^{2M} a_s(i)p_{2,i}(x) + u(x,t_s) - u(0,t_s)\\
&\quad + x[u'(0,t) - u'(0,t_s)] + u(0,t)\\
\dot{u}(x,t) &= \sum_{i=1}^{2M} a_s(i)p_{2,i}(x) + \dot{u}(0,t) + x\dot{u}'(0,t).
\end{aligned} \tag{6.17}
$$

On the grounds of the boundary conditions we have $u(0,t) = u(0,t_s) = \dot{u}(0,t) = 0$. The conditions $u(1,t) = \dot{u}(1,t) = 0$ give

$$
\begin{aligned}
u'(0,t) - u'(0,t_s) &= -(t-t_s)\sum_{i=1}^{2M} a_s(i)q(i)\\
\dot{u}'(0,t) &= -\sum_{i=1}^{2M} a_s(i)q(i),\ q(i) = p_{2,i}(1).
\end{aligned} \tag{6.18}
$$

It follows from (2.6), (2.7) that

$$
q(i) = \begin{cases} 0.5 \text{ if } i = 1,\\ \frac{1}{4m^2} \text{ if } i > 1. \end{cases} \tag{6.19}
$$

Substituting (6.18) into (6.17) and discretizising the results by assuming $x \to x_l,\ \ t \to t_{s+1}$ we obtain

$$
\begin{aligned}
u''(l,s+1) &= \Delta t a_s(:)H(:,l) + u''(l,s)\\
u'(l,s+1) &= \Delta t a_s(:)[P_1(:,l) - E(l)q(:)] + u'(l,s)\\
u(l,s+1) &= \Delta t a_s(:)[P_2(:,l) - x(l)q(:)] + u(l,s)\\
\dot{u}(l,s+1) &= a_s(:)[P_2(:,l) - x(l)q(:)].
\end{aligned} \tag{6.20}
$$

In these equations $E(l)$ denotes the $2M$-dimensional unit vector. To simplify the writing of the formulas, the matrix formulation is used and the notations $\Delta t = t_{s+1} - t_s,\ u(l,s) = u(x_l,t_s)$ etc. are used. There are several possibilities for treating the nonlinearity in (6.15). In the following the scheme,

$$
\dot{u}(l,s+1) = -\frac{1}{2}u(l,s)u'(l,s)u'(l,s) + \nu u''(l,s), \tag{6.21}
$$

which leads us from the time layer, t_s to t_{s+1} is used.

Substituting (6.20) into (6.21) we obtain

$$a_s(:)[P_2(:,l) - q(:)x(l)] = -0.5u(l,s)u'(l,s) - \nu u''(l,s). \tag{6.22}$$

From (6.22) the wavelet coefficients a_s can be successively calculated. This process is started with

$$\begin{aligned} u(l,0) &= \sin[2\pi x(l)] \\ u'(l,0) &= 2\pi \cos[2\pi x(l)] \\ u''(l,0) &= -4\pi^2 \sin[2\pi x(l)] \end{aligned} \tag{6.23}$$

In the case of a small viscosity ν, the solution develops into a saw-tooth wave and at $x = 0.5$ a shock develops. For estimating the efficiency of the solution, it is expedient to calculate the maximum value of the gradient u' at $x = 0.5$. According to the analytical solution [3, 10] for $\nu = (400\pi)^{-1}$, the theoretical maximum is $u' = 304.0$ and takes place at $t_{max} = 0.51$.

Computer simulation was carried out for $\nu = (400\pi)^{-1}$. Results for $J = 5$ (32 collocation points), $\Delta t = 0.001$, $t = 0.35$ are plotted in Fig. 6.1a. It follows from this figure that in spite of the small number of the collocation points the solution describes quite well the sawtooth effect. With increasing t oscillations near the point $x = 0.5$ appears (Fig. 6.1b). For getting more exact results, the value of J must be increased. Results for $J = 7$, $t = 0.51$ are plotted in Fig. 6.1c. In this case $\max \mid \partial u/\partial x \mid = 400.4$. That number is different from the analytical value 304.0, but this fact seems to be nonessential: if we calculate the angle $\delta = \arctan(\max |\partial u/\partial x|)$, then our solution gives $\delta = 89.86°$, while in the case of the analytic solution we have $\delta = 89.81°$.

Results of numerical solution for the Burgers equation, with the aid of other methods, can be found in [3, 10, 32]. Comparison with these algorithms shows that the Haar wavelet method is competitive and efficient. The advantages of our method are its simplicity and speed of convergence, which results from the sparseness of the transform matrices H, P_1, P_2. As it follows from Fig. 6.1d, the number of significant wavelet coefficients is quite small, while in the case of other wavelet methods it can be up to 200–400 [5].

From this section one more good feature of the Haar wavelets becomes evident-their possibility to detect singularities and a simple way of treating them.

6.4 Sine-Gordon Equation

In this section the classical form of the sine-Gordon equation

$$\frac{\partial^2 u}{\partial x_*^2} - \frac{\partial^2 u}{\partial t_*^2} = \sin u, \quad x_* \in [x_{in}, x_{fin}], \quad t_* \geq t_{in} \tag{6.24}$$

is considered.

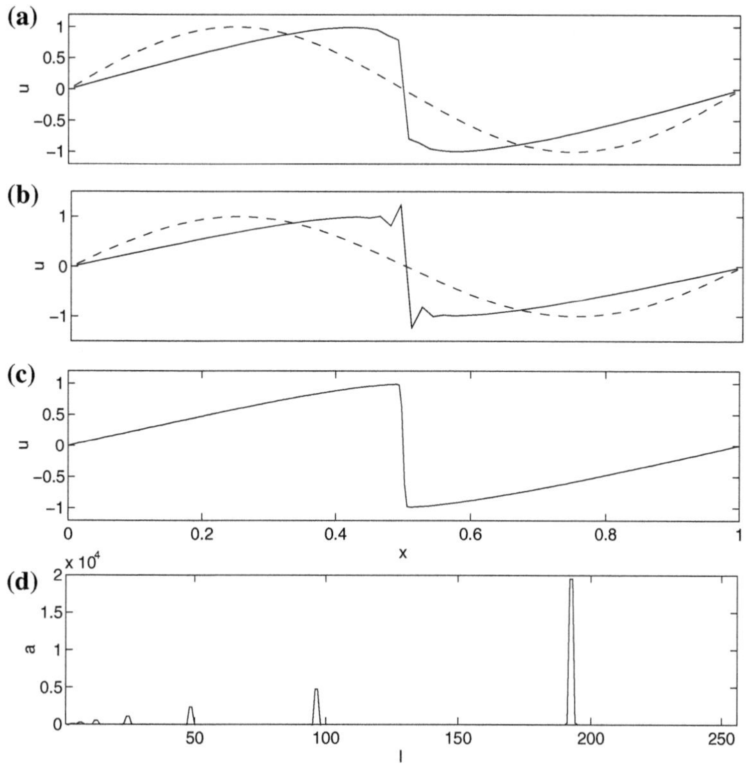

Fig. 6.1 Solutions of the Burgers equation for $\nu = (400\pi)^{-1}$, $\Delta t = 0.001$: **a** solution for $J = 5$, $t = 0.35$; **b** $J = 5$, $t = 0.37$ (*dashed line* denotes the solution for $t = 0$); **c** $J = 7$, $t = 0.51$; **d** wavelet coefficients for the case *c* (l is the queue number of collocation points)

Changing the variables

$$x = \frac{1}{L}(x_* - x_{in}), \quad t = t_* - t_{in}, \quad L = x_{fin} - x_{in}$$

we obtain

$$\frac{1}{L^2}u'' - \ddot{u} = \sin u, \quad x \in [0, 1], \quad t \geq 0. \tag{6.25}$$

This equation has an analytical solitary wave solution

$$u(x, t) = 4 \arctan[\exp(z)], \tag{6.26}$$

where

$$z = \alpha(x - \beta t), \ \alpha = \frac{L}{\sqrt{1 - L^2\beta^2}}. \tag{6.27}$$

To (6.25) belong the initial and boundary conditions

$$\begin{aligned} u(0,t) &= \varphi(t), & u'(0,t) &= \psi(t) \\ u(x,0) &= f(x), & \dot{u}(x,0) &= g(x)' \end{aligned} \tag{6.28}$$

where φ, ψ, f, g are prescribed functions.

Numerical solution of the Sine-Gordon equation is discussed in many papers; mostly the conventional approaches such as the finite difference method, spectral methods of integration etc. are applied. From such papers we cite here Ablowich et al. [1], Forinash and Willis [13], Conzalez et al. [14], Lu and Schmid [28].

Next we present the Haar wavelet solution, which is sought in the form

$$\ddot{u}''(x,t) = \sum_{i=1}^{2M} a_s(i)h_i(x), \quad t \in [t_s, t_{s+1}], \quad x \in [0,1]. \tag{6.29}$$

This equation is integrated twice in regard to x in the limits $[0, x]$ and in regard to t in the limits $[t_s, t_{s+1}]$. By doing this we obtain

$$\begin{aligned} \dot{u}''(x,t) &= (t-t_s)\sum_{i=1}^{2M} a_s(i)h_i(x) + \dot{u}''(x,t_s) \\ u''(x,t) &= \frac{1}{2}(t-t_s)^2\sum_{i=1}^{2M} a_s(i)h_i(x) + (t-t_s)\dot{u}''(x,t_s) + u''(x,t_s) \\ \ddot{u}(x,t) &= \sum_{i=1}^{2M} a_s(i)p_{2,i}(x) + x\ddot{u}'(0,t) + \ddot{u}(0,t) \\ \dot{u}(x,t) &= (t-t_s)\sum_{i=1}^{2M} a_s(i)p_{2,i}(x) + \dot{u}(x,t_s) \\ &\quad + x[\dot{u}'(0,t) - \dot{u}'(0,t_s)] + \dot{u}(0,t) - \dot{u}(0,t_s) \\ u(x,t) &= \frac{1}{2}(t-t_s)^2\sum_{i=1}^{2M} a_s(i)p_{2,i}(x) + u(x,t_s) \\ &\quad + x[u'(0,t) - u'(0,t_s) - (t-t_s)\dot{u}'(0,t_s)] \\ &\quad + u(0,t) - u(0,t_s) - (t-t_s)\dot{u}(0,t_s) \end{aligned} \tag{6.30}$$

These results are discretized by replacing $x \longrightarrow x_l,\ t \longrightarrow t_{s+1}$. Again the notations $\Delta t = t_{s+1} - t_s,\ u(l,s) = u(x_l, t_s)$ etc. are used. Taking into account the initial conditions (6.28) the Eq. (6.30) can be put into the form

$$
\begin{aligned}
\dot{u}''(l, s+1) &= \Delta t a_s(:) H(:, l) + \dot{u}''(l, s) \\
u''(l, s+1) &= \frac{1}{2}\Delta t^2 a_s(:) H(:, l) + u''(l, s) + \Delta t \dot{u}''(l, s) \\
\ddot{u}(l, s+1) &= a_s(:) P_2(:, l) + \ddot{\varphi}(s+1) + x_l \ddot{\psi}(s+1) \\
\dot{u}(l, s+1) &= \Delta t a_s(:) P_2(:, l) + \dot{u}(s, l) + \dot{\varphi}(s+1) - \dot{\varphi}(1) \\
&\quad + x_l[\dot{\psi}(s+1) - \dot{\psi}(s)] \\
u(l, s+1) &= \frac{1}{2}\Delta t^2 a_s(:) P_2(:, l) + u(l, s) + \Delta t \dot{u}(l, s) + \varphi(s+1) \\
&\quad - \varphi(s) - \Delta t \dot{\varphi}(s) + x_l[\psi(s+1) - \psi(s) - \Delta t \dot{\psi}(s)]. \qquad (6.31)
\end{aligned}
$$

By replacing these results into (6.25) we get a linear matrix equation for calculating the wavelet coefficients $a_s(:)$:

$$
a_s(:) P_2(:, l) = \frac{1}{L^2} u''(s, l) + \sin u(s, l) - x_l \ddot{\varphi}(s+1) - \ddot{\psi}(s+1). \qquad (6.32)
$$

Now let us specify the initial and boundary conditions. If we want to get the classical solitary wave solution we must take into account (6.26):

$$
\begin{aligned}
f(x) &= u(x, 0) = 4 \arctan[\exp(\alpha x)] \\
g(x) &= u'(x, 0) = \alpha V(\alpha x) \\
\varphi(t) &= u(0, t) = 4 \arctan[\exp(-\alpha \beta t)] \\
\psi(t) &= \dot{u}(0, t) = -\alpha \beta V(-\alpha \beta t) \qquad (6.33)
\end{aligned}
$$

where $V(z) = \frac{4e^z}{1+e^{2z}}$.

For solving (6.32) we need in addition the functions $u''(x, 0)$, $\dot{u}''(x, 0)$ $\ddot{\varphi}(t)$, $\ddot{\varphi}\,'(t)$, $\varphi'(t)$, $\dot{\varphi}'(t)$, which can be calculated by differentiating Eq. (6.33).

Computer simulation was carried out for $t_{in} = 10$, $t_{fin} = 30$, $L = 20$, $\beta = 0.025$. The computed results were compared with the exact solution (6.26). The accuracy of our approach was estimated by the error function

$$
v(t) = \frac{1}{2M} \parallel u(x, t) - u_{ex}(x, t) \parallel = \frac{1}{2M} \left\{ \sum_{i=1}^{2M} [u(x_i, t) - u_{ex}(x_i, t)]^2 \right\}^{\frac{1}{2}}. \qquad (6.34)
$$

The calculations showed that the function $v(t)$ increases monotonically; therefore for the error estimate is taken $v(t_{max})$. Some results of computation are presented in the Table 6.1 and in Fig. 6.2.

It follows from this table that already in the cases $J = 4$ or $J = 5$ we get results which visually coincide with the exact solution. For the time step was taken $\Delta t = 0.005$ or $\Delta t = 0.001$; further diminution of Δt did not give any essential effect. It can be seen from Fig. 6.2c that the error v distribution along the x-axis is not uniform (as it was to be awaited the biggest values of v appear in the large gradient regions).

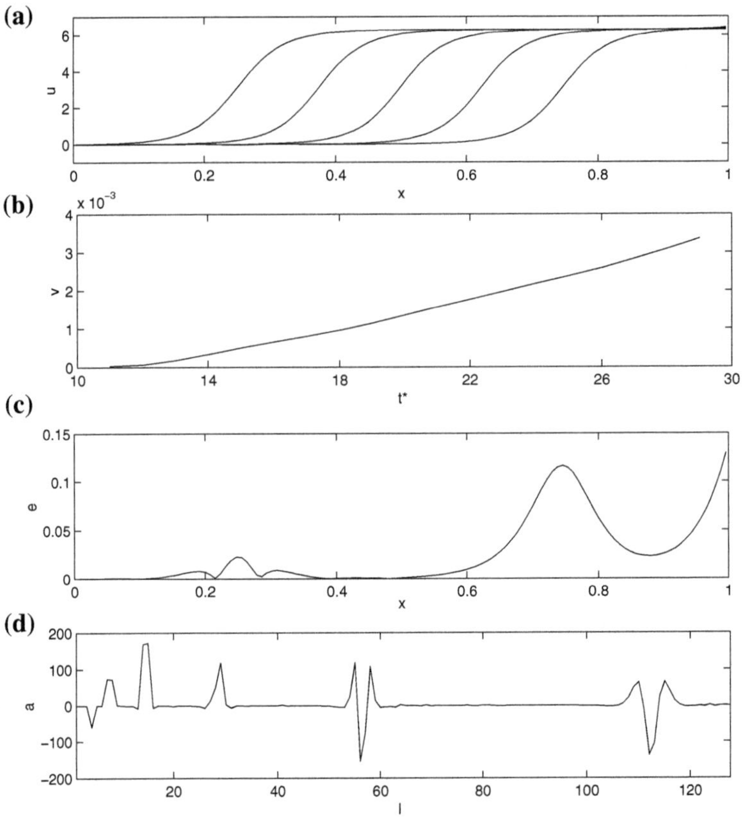

Fig. 6.2 Solution of the sine-Gordon equation for $J = 6$, $\Delta t = 0.001$: **a** solution for the instants $t_* = 10 + 5i$ $i = 0, 1, 2, 3, 4$; **b** error function $v(t_*)$; **c** error function $e(x) = | u(x, t_{max}) - u_{lx}(x, t_{max}) |$; **d** wavelet coefficients at $t_* = 30$

Table 6.1 Error estimates of Eq. (6.15)

J	$2M$	$v(t_{max})$	
		$\Delta t = 0.05$	$\Delta t = 0.01$
4	16	0.051	0.038
5	32	0.018	0.009
6	64	0.0096	0.0036

According to Fig. 6.2d the number of significant wavelet coefficients is again quite small.

The results of this section demonstrate the efficiency of the Haar method for investigating the propagation of solitary waves.

6.5 Related Papers

Wavelet based methods have been applied for solving PDEs from beginning of the early 1990s. In the last two decades this problem has aroused attention and numerous papers on this topic have been published. Due to the great number of such papers we cannot give an overview of them all, but confine ourselves to some comments. From earlier results we cite here the papers by Dahmen et al. [11], Engquist et al. [12], Bertoluzza and Naldi [4], Comincioli et al. [10].

In the case of evolution equations, often singularities and steep gradients in a small region of the analyzed function appear. In this case, adaptive algorithms with the refinement of the local grid in the areas of singularities are needed. From such papers we refer here [2, 9, 30, 31]. A shortcoming of these methods is the fact that computation of complicated connection coefficients is required.

The first paper in which the Haar method for solving PDEs was applied belongs to Chen and Hsiao [8]; in this paper, the diffusion equation and telegraph equation are solved.

In the subsequent contributions, the Haar method is applied for solution of different PDEs. The heat equation was discussed by Cattani [6]. The same equation was applied for estimating depth profile of soil temperature by Hariharan and Kannan [16]. Nonlinear parabolic equations were solved by Hariharan and Kannan [20]. The one-dimensional Burger equation was solved by Jiwari [23], an efficient numerical scheme based on the quasilinearization process is developed.

The Fisher's equation

$$\dot{u} = u'' + u(1 - u), \tag{6.35}$$

which has important applications to chemistry and biology was solved by Hariharan et al. [22].

The generalized Burgers-Huxley equation

$$\dot{u} + \alpha u^{\delta} u' - u'' = \beta u(1 - u^{\delta})(u^{\delta} - \gamma) \tag{6.36}$$

was analysed with the aid of the Haar method by Çelik [7].

This equation has some important special forms. When $\alpha = 0,\ \delta = 1$ it is reduced to the Huxley equation, which describes nerve pulse propagation in nerve fibers. At $\delta = 1$ and $\alpha \neq 0,\ \beta \neq 0$ Eq. (6.36) is turned into the Burgers-Huxley equation which is important for describing the convection effects and diffusion transport. When $\alpha = 0,\ \beta = 1,\ \delta = 1$ Eq. (6.36) becomes the Fitz Hugh-Nagumo equation which is used in the circuit theory, biology and population genetics; this equation was solved making use of Haar wavelets in [19].

All these equations belong to the group of nonlinear convection-diffusion equations, which are discussed by Çelik [7], Hariharan and Kannan [18, 21].

The Cohn-Allen equation

$$\ddot{u} = u'' + u^3 + u, \tag{6.37}$$

which has many applications in biology, quantum mechanics and plasma physics, was solved in [17].

The Klein-Gordon equation

$$\ddot{u} + \alpha u'' + \gamma u^{\nu} = f(x,t), \;\; \nu = 2 \; or \; \nu = 3 \tag{6.38}$$

has attracted much attention in studying solitons and quantum mechanics. The Haar method for its solution was proposed in [15]. The cubic-spline method was applied in [24].

A numerical method for solving PDE called "the weak formulation based on Haar wavelet method" was proposed by Majak et al. [29] and applied for solving the Burgers equation

$$\dot{u} + uu' = \varepsilon u''. \tag{6.39}$$

This equation was discretized in time by the forward Euler method:

$$u_{j+1} = u_j + \Delta t(-u_j u'_j + \epsilon u''_j). \tag{6.40}$$

The weak form of this equation is

$$\int_0^1 \nu(\tau)[-u_{j+1} + u_j + \Delta t u_j u'_j + \epsilon u''_j]d\tau = 0, \;\; j = 1, \ldots, n, \tag{6.41}$$

where ν stands for the weight function. The functions ν and u_j are expanded into the Haar series

$$u''_j = aH, \;\; \nu'' = bH. \tag{6.42}$$

The unknown coefficients a and b are calculated from (6.41). A numerical example is given.

References

1. Ablowitz, M., Herbst, B., Schober, C.: On the numerical solution of the sine-Gordon equation, ii. Performance of numerical schemes. J. Comp. Phys. **131**, 354–367 (1997)
2. Alam, J., Kevlahan, N.K., Vasilyev, O.: Simultaneous space time adaptive wavelet solution of nonlinear parabolic differential equations. J. Comput. Phys. **214**, 829–857 (2006)
3. Basdevant, C., Deville, M., Haldenwang, P., Lacroix, J., Ouazzani, J., Peyret, R., Orlandi, P., Patera, A.: Spectral and finite difference solutions of the Burgers equation. Comput. Fluids **14**, 23–41 (1986)
4. Bertoluzza, S., Naldi, G.: A wavelet collocation method for the numerical solution of partial differential equations. Appl. Comput. Harmon. Anal. **3**, 1–9 (1996)
5. Beylkin, G., Kaiser, J.: An adaptive pseudo-wavelet approach for solving nonlinear partial differential equations. In: W. Dahmen, A.J. Kurdila, F. Ostwalds (eds.) Multiscale Wavelet Methods for PDE, pp. 137–197. Academic press, San Diego (1997)
6. Cattani, C.: Haar wavelets based technique in evolution problems. Proc. Est. Acad. Sci. Phys. **53**, 45–63 (2004)

7. Çelik, I.: Haar wavelet method for solving generalized Burgers-Huxley equation. Arab J. Math. Sci. **18**, 25–37 (2012)
8. Chen, C., Hsiao, C.: Haar wavelet method for solving lumped and distributed-parameter systems. IEEE Proc. Control Theory Appl. **144**, 87–94 (1997)
9. Chiavassa, G., Guichaoua, M., Liandrat, J.: Two adaptive wavelet algorithms for non-linear parabolic partial differential equations. Comput. Fluids **31**, 467–480 (2002)
10. Comincioli, V., Naldi, G., Scapolla, T.: A wavelet-based method for numerical solution of nonlinear evolution equations. Appl Numer Math **33**, 291–297 (2000)
11. Dahmen, W., Kurdila, A., Oswald, P.: Multiscale wavelet methods for partial differential equations. Academic Press, London (1997)
12. Engquist, B., Osher, S., Zhong, S.: Fast wavelet based algorithms for linear evolution equations. SIAM J. Sci. Comput. **15**, 755–775 (1994)
13. Forinash, K., Willis, C.: Nonlinear response of the sine-Gordon breather to an a.c. driver. Physica D **149**, 95–106 (2001)
14. González, J., Bellorin, A., Guerrero, L.: How to excite the internal modes of sine-Gordon solitons. Chaos, Solitons and Fractals **17**, 907–919 (2003)
15. Hariharan, G.: Haar wavelet method for solving the Klein-Gordon and the sine-Gordon equations. Int. J. Nonlinear Sci. **11**, 180–189 (2011)
16. Hariharan, G., Kannan, K.: Haar wavelet in estimating depth profile of soil temperature. Appl. Math. Comput. **210**, 119–125 (2009)
17. Hariharan, G., Kannan, K.: Haar wavelet method for solving Cahn-Allen equation. Appl. Math. Sci. **3**, 2523–2533 (2009)
18. Hariharan, G., Kannan, K.: A comparison of Haar wavelet and Adomain decomposition method for solving one-dimensional reaction-diffusion equations. Int. J. Appl. Math. Comput. **2**, 50–61 (2010)
19. Hariharan, G., Kannan, K.: Haar wavelet method for solving Fitz-Hugh-Nagumo equation. Int. J. Math. Stat. Sci **2**, 2 (2010)
20. Hariharan, G., Kannan, K.: Haar wavelet method for solving some nonlinear parabolic equations. J. Math. Chem. **48**, 1044–1061 (2010)
21. Hariharan, G., Kannan, K.: A comparative study of Haar wavelet method and homotopy pertubation method for solving one dimensional diffusion equations. Int. J. Math. Comput. **3**, 21–34 (2011)
22. Hariharan, G., Kannan, K., Sharma, K.: Haar wavelet method for solving Fisher's equation. Appl. Math. Comput. **211**, 284–292 (2009
23. Jiwari, R.: A Haar wavelet quasilinearization approach for numerical simulation of Burgers' equation. Comput. Phys. Commun. **183**, 2413–2423 (2012)
24. Lakestani, M., Dehghan, M.: Collocation and finite difference-collocation methods for the solution of nonlinear Klein-Gordon equation. Comput. Phys. Commun. **181**, 1392–1401 (2010)
25. Lepik, Ü.: Numerical solution of differential equations using Haar wavelets. Math. Comput. in Simul. **68**, 127–143 (2005)
26. Lepik, Ü.: Application of the Haar wavelet transform to solving integral and differential equations. Proc. Estonian Acad. Sci. Phys. Math. **56**, 28–46 (2007)
27. Lepik, Ü.: Numerical solution of evolution equations by the Haar wavelet method. Appl. Math. Comp. **185**, 695–704 (2007)
28. Lu, X., Schmid, R.: Sympletic integration of sine-Gordon type systems. Math. Comput. Simul. **50**, 255–263 (1999)
29. Majak, J., Pohlak, M., Eerme, M., Lepikult, T.: Weak formulation based Haar wavelet method for solving differential equations. Appl. Math. Comput. **211**, 488–494 (2009)
30. Sandeep, K., Gaur, S., Dutta, D., Kushwaha, H.: Wavelet based schemes for linear advection dispersion equation. Appl. Math. Comput. **218**, 3786–3798 (2011)
31. Vasilyev, O., Paolucci, S.: A dynamically adaptive multilevel wavelet collocation method for solving partial differential equations in a finite domain. Journal Comp. Phys. **125**, 498–512 (1996)
32. Vasilyev, O., Paolucci, S., Sen, M.: A multilevel wavelet collocation method for solving partial differential equations in finite domain. J Comp Phys **120**, 33–47 (1995)

Chapter 7
Solving PDEs with the Aid of Two-Dimensional Haar Wavelets

7.1 Problem Statement and Method of Solution

Consider linear PDE

$$\sum_{\gamma=0}^{\Gamma}\sum_{\lambda=0}^{\Lambda} D_{\gamma\lambda}\frac{\partial^{(\gamma+\lambda)}u}{\partial x^{\gamma}\partial y^{\lambda}} = f(x,y), \tag{7.1}$$

where Γ, Λ are given constants and $D_{\gamma\lambda}$, f are prescribed functions. The independent variables x, y belong to a domain Σ, which has the boundary σ. We have to calculate the function $u(x, y)$, which satisfies the required boundary conditions. For simplicity sake, we confine ourselves to problems, where the domain Σ is a rectangle $x \in [0, L_1]$, $y \in [0, L_2]$ and divide the intervals $[0, L_1]$, $[0, L_2]$ into $2M_1$ and $2M_2$ parts of equal length, respectively. According to the Haar wavelet method the solution is sought in the form

$$\frac{\partial^{(\Gamma+\Lambda)}u}{\partial x^{\Gamma}\partial y^{\Lambda}} = \sum_{i=1}^{2M_1}\sum_{l=1}^{2M_2} a_{il}h_i(x)h_l(y), \tag{7.2}$$

where a_{il} are the wavelet coefficients and $h_i(x)$, $h_l(y)$ the Haar functions. By multiple integrating (7.2), the lower order derivatives and the function u itself are calculated. In this procedure, unknown functions $\varphi_1(x), \varphi_2(x), \ldots, \varphi_{\Gamma-1}(x), \psi_1(y), \psi_2(y), \ldots, \psi_{\Lambda-1}(y)$ appear; they are calculated from the boundary conditions. We satisfy (7.1) in the collocation points x_r,y_s, where $r \in [1, 2M_1]$, $s \in [1, 2M_2]$. By doing this we get the following system of linear equations

$$\sum_{i=1}^{2M_1}\sum_{l=1}^{2M_2} a_{il}R_{ilrs} = f(x_r, y_s), \tag{7.3}$$

from which the wavelet coefficients a_{il} can be calculated (the symbol R_{ilrs} denotes a function which will be specified later on). Since we do not possess algorithms for

Ü. Lepik and H. Hein, *Haar Wavelets*, Mathematical Engineering,
DOI: 10.1007/978-3-319-04295-4_7,

dealing with the fourth-order matrices we must transform the system into a form, where only two-order matrices appear. This can be done by introducing new indices

$$\alpha = 2M_1(i-1)+l, \quad \beta = 2M_2(r-1)+s \tag{7.4}$$

Denoting $a_{il} \to b(\alpha), f(x_r, y_s) \to F(\beta), R_{ijrs} \to S(\alpha, \beta)$ and we can rewrite (7.3) in form

$$\sum_{\alpha=1}^{2M_1}\sum_{\beta=1}^{2M_2} b(\alpha)S(\alpha, \beta) = F(\beta). \tag{7.5}$$

Here b and F are $2M_1 * 2M_2$ dimensional row vectors and S is a $(2M_1)^2 * (2M_2)^2$ dimensional matrix. It is expedient to put (7.5) into the matrix form

$$bS = F. \tag{7.6}$$

After evaluating the coefficients $b(\alpha)$ from (7.5) it is not difficult to restore the original matrix of the wavelet coefficients matrix a_{il}. According to (7.4), we have $\alpha/(2M_1) = i-1+l/(2M_1), \beta/2M_2 = r-1+s/2M_2$. Consequently, the integer part of these expressions gives $i-1$ and $r-1$. The remainders of these divisions are equal to l and s. Let us take a numerical example. Assume that $M_1 = M_2 = 4$ and consider the term $b(10)S(10, 25)$. Since $10/8 = 1+remainder(2)$ and $25/8 = 3+remainder(1)$ we have $i = 2,\ l = 2,\ r = 4,\ s = 1$ and $b(10),\ S(10, 25) \to a_{24}R_{2441}$.

By integrating (7.2) Γ-times in regard to x and Λ-times in regard to y, we obtain

$$u(x, y) = \sum_{i=1}^{2M_1}\sum_{l=1}^{2M_2} a_{il}p_\Gamma(x)p_\Lambda(y) + \Psi(x, y). \tag{7.7}$$

In this formula the integrals $p_\Gamma(x)$ and $p_\Lambda(y)$ are calculated according to (7.6) and (7.7) the function $\Psi(x, y)$ incorporates the functions $\varphi_1(x), \varphi_2(x), \ldots, \psi_1(y), \psi_2(y)$, $\ldots$ appearing in the course of integrations of (7.2). Details of this method are explained by solving the two problems in Sects. 7.2 and 7.3.

7.2 Diffusion Equation

Consider once more the diffusion equation

$$\frac{\partial u}{\partial t} = A\frac{\partial^2 u}{\partial x^2}, \quad (x, t) \in [0, 1] \tag{7.8}$$

for the initial condition $u(x, 0) = g(x)$ and boundary conditions $u(0, t) = u(1, t) = 0$.

Let us confine to the case $M_1 = M_2 = M$. The two-dimensional Haar wavelet solution is sought in the form

$$\frac{\partial^3 u}{\partial t \partial x^2} = \sum_{i=1}^{2M} \sum_{l=1}^{2M} a_{il} h_i(x) h_l(t). \tag{7.9}$$

By multiple integration and considering the initial condition we find

$$\begin{aligned}\frac{\partial^2 u}{\partial x^2} &= \sum_i \sum_l a_{il} h_i(x) p_{1l}(t) + \frac{\partial^2 g}{\partial x^2} \\ \frac{\partial u}{\partial t} &= \sum_i \sum_l a_{il} p_{2i}(x) h_l(t) + x \frac{\partial^2 u}{\partial t \partial x} |_{x=0} + \frac{\partial u}{\partial t} |_{x=0}\end{aligned} \tag{7.10}$$

For simplicity sake here and in the following, the limits of summation are not written out. In view of the boundary conditions we have

$$\frac{\partial u}{\partial t} |_{x=0} = \frac{\partial u}{\partial t} |_{x=1} = 0. \tag{7.11}$$

It follows for $x = 1$ from (7.10)

$$\frac{\partial^2 u}{\partial t \partial x} |_{x=0} = -\sum_i \sum_l a_{il} q_2(i) h_l(t), \; q_2(i) = p_{2,i}(1). \tag{7.12}$$

Replacing this result back into (7.10) we obtain

$$\frac{\partial u}{\partial t} = \sum_i \sum_l a_{il} [p_{2i}(x) - x q_2(i)] h_l(t). \tag{7.13}$$

These results are substituted into (7.8); satisfying the obtained equation in the collocation points x_r, y_s we get the system

$$\sum_i \sum_l a_{il} R_{ilrs} = g''(x_r), \tag{7.14}$$

where

$$R_{ilrs} = [P_2(i, r) - x_r q_2(i)] H(l, s) - A\, H(i, l) P_1(r, s). \tag{7.15}$$

Introducing the indices α, β according to (7.4) this result obtains the form (7.5) or (7.6). The wavelet coefficients a_{ij} are calculated in the way indicated in Sect. 7.1. For evaluating the function $u(x, y)$ we integrate (7.13) with regard to t and obtain

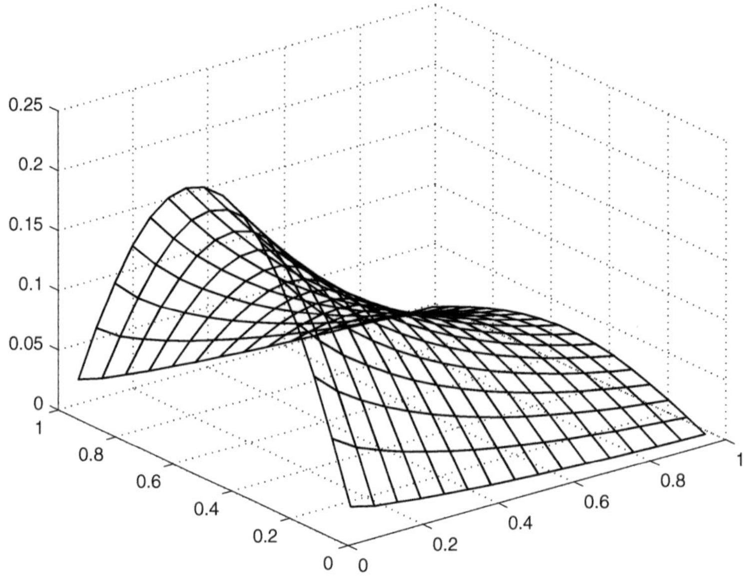

Fig. 7.1 Solution of the diffusion equation for $J = 3$

$$u(x,t) = \sum_i \sum_l a_{il}[p_{2i}(x) - xq_2(i)]p_{1l}(t) + g(x) \tag{7.16}$$

Computer simulation was carried out for $A = 0.2$ and $g(x) = x(1 - x)$ For J the values 2, 3, 4 were taken. The solution for $J = 3$ is plotted in Fig. 7.1. These results were compared with the Fourier solution (7.13).

For estimating the exactness of the wavelet solution (7.16), the difference between both solutions in the collocation point $[x(M), y(M)]$ was calculated and we found that $u[x(M), y(M)]$ was $5.4E - 4$ for $J = 2$, $1.4E - 4$ for $J = 3$ and $3.5E - 5$ for $J = 4$. So we state that already for $J = 2$ (64 collocation points) the accuracy of the Haar wavelet solution is rather good.

7.3 Poisson Equation

Consider the Poisson equation

$$\frac{\partial^2 u}{\partial x^2} + \frac{\partial^2 u}{\partial y^2} = f(x,y) \tag{7.17}$$

on a square $x \in [0, 1]$, $y \in [0, 1]$ with the boundary conditions $u(x, 0) = u(0, y) = u(x, 1) = 0,\ u(1, y) = g(y)$.

The solution is started by assuming

$$\frac{\partial^4 u}{\partial x^2 \partial y^2} = \sum_i \sum_l a_{il} h_i(x) h_l(y), \tag{7.18}$$

By integrating this equation twice with regard to x and twice with regard to y we find

$$\begin{aligned} \frac{\partial^2 u}{\partial x^2} &= \sum_i \sum_l a_{il} h_i(x) p_{2l}(y) + y\varphi_1''(x) + \varphi_2''(x) \\ \frac{\partial^2 u}{\partial y^2} &= \sum_i \sum_l a_{il} p_{2l}(x) h_l(y) + x\psi_1''(y) + \psi_2''(y) \end{aligned} \tag{7.19}$$

By multiple integrating we obtain

$$u(x, y) = \sum_i \sum_l a_{il} p_{2i}(x) p_{2l}(y) + y\varphi_1(x) + \varphi_2(x) + x\psi_1(y) + \psi_2(y) \tag{7.20}$$

Next the boundary conditions are satisfied:

(i) It follows from $u(x, 0) = 0$ that $\varphi_2(x) = -x\psi_1(0) - \psi_2(0)$.
(ii) The condition $u(0, y) = 0$ gives $\psi_2(y) = -y\varphi_1(0) - \varphi_2(0)$.
(iii) Satisfying the condition $u(x, 1) = 0$ we obtain $\varphi_1(x) = -\sum_i \sum_l a_{il} p_{2i}(x) q_2(l) + \varphi_1(0) - x[\psi 1(1) - \psi_1(0)]$.
(iv) In view of $u(1, y) = g(y)$ we find $\psi_1(y) = -\sum_i \sum_l a_{il} q_2(i)[-p_{2l}(y) + yq_2(l)] + \psi_1(0) + y[\psi_1(1) - \psi_1(0)] + g(y)$. Now (7.19) obtains the form

$$\begin{aligned} \frac{\partial^2 u}{\partial x^2} &= \sum_i \sum_l a_{il} h_i(x)[p_{2l}(y) - yq_2(l)] \\ \frac{\partial^2 u}{\partial y^2} &= \sum_i \sum_l a_{il} [p_{2l}](x) - xq_2(i)] h_l(y) + xg''(y) \end{aligned} \tag{7.21}$$

Substituting this result into (7.17) and satisfying the obtained equation in the collocation points we find

$$\sum_i \sum_l a_{il} R_{ilrs} = F(r, s) \tag{7.22}$$

where

$$\begin{aligned} R_{ilrs} &= H(i, r) P_2(l, s) + P_2(i, r) H(l, s) - H(i, r) q_2(l) y_s - q_2(i) x_r) H(l, s) \\ F(r, s) &= f(x_r, y_s) - x_r g''(y_s). \end{aligned} \tag{7.23}$$

For solving this system again, the technique of Sect. 7.1 is applied. The final result is

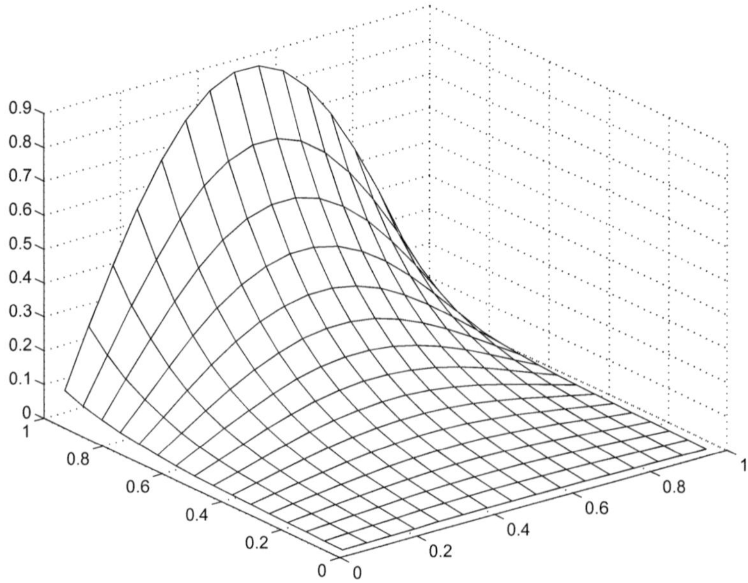

Fig. 7.2 Solution of the Poisson equation for $J = 3$

$$u(x, y) = \sum_i \sum_l a_{il}\{p_{2i}(x)[p_{2l}(y) - yq_2(l)] + xq_2(i)[-p_{2l}(y) + yq_2(l)]\} + g(y) \tag{7.24}$$

Example 7.1: Solve (7.17) for $g = 0$ and

$$f(x, y) = 6xy[y^2(1 - 2x)(1 - y) - x^2(1 - 2y)(1 - x)]. \tag{7.25}$$

The exact solution for this case is

$$u_{ex} = x^3y^3(1 - x)(1 - y). \tag{7.26}$$

For estimating the accuracy of the wavelet results, the matrix norm from the MATLAB library $\Delta(\eta) = norm(u - u_{ex}, \eta)$ is used; the case $\eta = 2$ gives the largest column sum, the case $\eta \to$ inf—the largest row sum. In the present case due to symmetry, both norms are equal and we can estimate the accuracy of our solution by the parameter $\delta = \Delta(2)/(2M)$.

Computer simulation gave the following error estimates: $\delta = 9.4E - 5$ for $J = 2$, $\delta = 2.4E - 5$ for $J = 3$, $\delta = 6.0E - 6$ for $J = 4$.

Example 7.2: Consider the case $f(x) = 3x^2$, $g(y) = A\sin(\pi y)$ The solution for $J = 3$ is plotted in Fig. 7.2. In the present case we do not know the exact solution, therefore we estimate the accuracy of the solution by the curves $u(x, 0.5)$ and $u(0.5, y)$, which are plotted in Fig. 7.3 for different values of J. It turned out that already for $J = 2$ and $J = 3$ these curves visually coincide. In addition to it, the

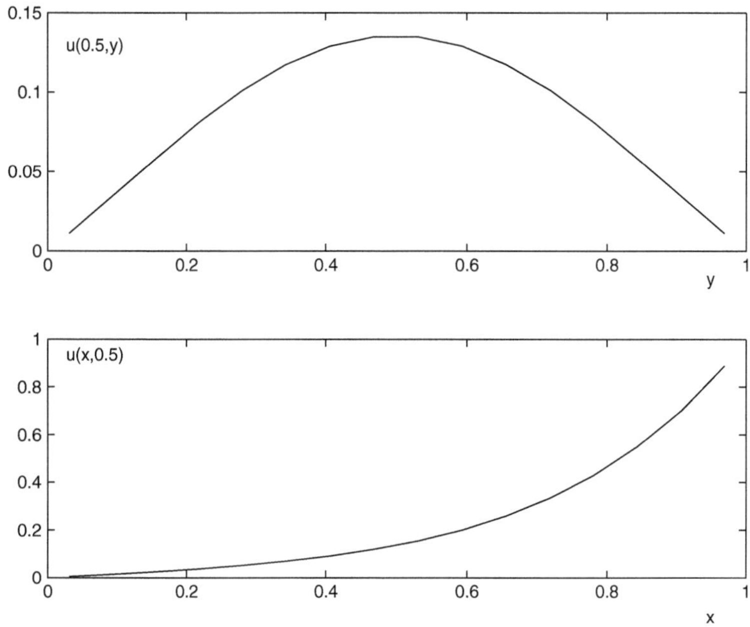

Fig. 7.3 Poisson equation: sections of the surface $u = u(x, y)$

value of $V = u(0.5, 0.5)$ was also computed and we found that $V = 0.13494$ for $J = 2$, $V = 0.13542$ for $J = 3$, $V = 0.13560$ for $J = 4$. So we can state that already the solution $J = 2$ (64 grid points) guarantees the necessary exactness in most cases.

7.4 Related Papers

There are numerous papers in which wavelet-oriented methods for solution of PDEs are applied. Mostly PDEs of mathematical physics (elliptic, parabolic and hyperbolic equations) are solved; usually the Daubechies wavelets are applied. Information about the earlier papers can be found from the book [7] and from the paper by Christon and Roach [6]. From the more recent papers we cite here the following. Bujurke et al. [3], Schwab and Stevenson [13] applied wavelet algorithms for solving boundary value problems for elliptic PDEs. Chen et al. [5] studied multiscale wavelet-based elements for adaptive finite element analysis. Zheng et al. [17] made use of discontinuous Legendre wavelets for solving elliptic PDEs. Chebyshev polynomials for biharmonic boundary value problems were applied by Mai-Duy and Tanner [11]. Coiflet wavelets for wavelet multiresolution analysis were used by Salazar and Lorduy [12]. Wavelet based analysis for two-dimensional advection-dispersion equation was carried out by Gaur et al. [8]. Navier-Stokes equations with the aid of divergence free wavelets were analysed by Zhou and He [18].

There are not many papers in which the Haar wavelet method was applied for solving PDEs. Abe et al. [1], Koro and Abe [9] applied Haar wavelets to time-domain Galerkin BEM. Lepik [10] presented a method of solution PDEs with the aid of the two-dimensional Haar wavelet; to demonstrate its efficiency the Poisson equation has been solved.

Solution of 2D and 3D Poisson equations and bilinear equations by the Haar wavelet method are reported by Shi et al. [14]. Numerical solution of elliptic boundary value problems with the aid of Haar and Legendre wavelets are discussed by Aziz et al. [2]. Numerical solution of boundary layer fluid flow problems was presented by Siraj-ul-Islam et al. [15]. Çelik [4] applied the Haar wavelet method for solving magneto-hydrodynamic flow problems.

Wang and Zhao [16] applied the Haar wavelet method for solving two-dimensional Burgers equations.

From the numerical results of all these papers, it follows that the Haar wavelet method for solving PDEs is simple, fast, reliable and with low computation costs.

In the present Chapter analysis of papers about fractional differential equations, optimal control theory and elastodynamics are missing. These problems will be discussed in the subsequent Chapters.

References

1. Abe, K., Koro, K., Itami, K.: An h-hierarchical Galerkin BEM using Haar wavelets. Eng. Anal. Bound. Elem. **25**, 581–591 (2001)
2. Aziz, I., Siraj-ul-Islam, Sarler, B.: Wavelets collocation methods for the numerical solution of elliptic BV problems. Aplied. Math. Model. **37**, 676–694 (2013)
3. Bujurke, N., Salimath, C., Kudenatti, R., Shiralashetti, S.: A fast wavelet-multigrid method to solve elliptic partial differential equations. Appl. Math. Comput. **185**, 667–680 (2007)
4. Çelik, I.: Haar wavelet approximation for magnetohydrodynamic flow equations. Appl. Math. Model. **37**, 3894–3902 (2013)
5. Chen, X., Xiang, J., Li, B., He, Z.: A study of multiscale wavelet-based elements for adaptive finite element analysis. Adv. Eng. Softw. **41**, 196–205 (2010)
6. Christon, M.A., Roach, D.W.: The numerical performance of wavelets for PDEs: the multi-scale finite element. Comput. Mech. **25**, 230–244 (2000)
7. Dahmen, W., Kurdila, A., Oswald, P.: Multiscale wavelet methods for partial differential equations. Academic Press, New York (1997)
8. Gaur, S., Singh, L., Singh, V., Singh, P.: Wavelet based multiscale scheme for two-dimensional advection-dispersion equation. Appl. Math. Model. **37**, 4023–4034 (2013)
9. Koro, K., Abe, K.: Application of Haar wavelets to time-domain BEM for the transient scalar wave equation. IOP Conf. Ser.: Mater. Sci. Eng. **10** (2010). doi:10.1088/1757-899X/10/1/012222.
10. Lepik, Ü.: Solving PDEs with the aid of two-dimensional Haar wavelets. Comput. Math. Appl. **61**, 1873–1879 (2011)
11. Mai-Duy, N., Tanner, R.: A spectral collocation method based on integrated Chebyshev polynomials for two-dimensional biharmonic boundary-value problems. J. Comput. Appl. Math. **201**, 30–47 (2007)

12. Salazar, A., Lorduy, G.H.: Approach to wavelet multiresolution analysis using Coiflets and a two-wave mixing arrangement. Optics Commun. **281**, 3091–3098 (2008)
13. Schwab, C., Stevenson, R.: Adaptive wavelet algorithms for elliptic PDEs on product domains. Math. Comput. **77**, 71–92 (2008)
14. Shi, Z., Cao, Y., Chen, Q.: Solving 2D and 3D Poisson equations and biharmonic equations by the Haar wavelet method. Appl. Math. Model. **36**, 5143–5161 (2012)
15. Siraj-ul-Islam, Sarler, B., Aziz, I., Fazal-i-Haq: Haar wavelet collocation method for the numerical solution of boundary layer fluid flow problems. Int. J. Thermal Sci. **50**, 686–697 (2011)
16. Wang, M., Zhao, F.: Haar wavelet method for solving two-dimensional Burgers equation. Adv. Intell. Soft Comput. **145**, 381–387 (2012)
17. Zheng, X., Yang, X., Su, H., Qiu, L.: Discontinuous Legendre wavelet element method for elliptic partial differential equations. Appl. Math. Comput. **218**, 3002–3018 (2011)
18. Zhou, X., He, Y.: Using divergence free wavelets for the numerical solution of the 2-D stationary Navier-Stokes equations. Appl. Math. Comput. **163**, 593–607 (2005)

Chapter 8
Fractional Calculus

8.1 Introduction

Although the conception of the fractional derivatives was already introduced in the middle of the nineteenth century by Riemann and Liouville, the first work, devoted exclusively to the subject of fractional calculus, is the book by Oldham and Spanier [21] published in 1974. After that the number of publications about the fractional calculus has rapidly increased. The reason for this is that some physical processes such as anomalous diffusion, complex viscoelasticity, behaviour of mechatronic and biological systems, rheology, etc. cannot be described adequately by the classical models.

At the present time, we possess several excellent monographs about fractional calculus, for example the book [14], in which there is also included a rather large and up-to-date Bibliography (928 items). Because of the enormous number of papers about this topic, we shall cite here only some papers which are closer to subject of this paper.

In a number of papers fractional differential equations are discussed; mostly these equations are transformed to fractional Volterra integral equations. For solution different techniques, such as Fourier and Laplace transforms, power spectral density, Adomian decomposition method, path integration, etc. are applied.

One-dimensional fractional harmonic oscillator is analyzed in [1, 2, 9, 20]. In [1, 9] the solution is obtained in terms of Mittag-Leffler functions using Laplace transforms; several cases of the forcing function equation are considered. In [9] the fractional equation of motion is solved by the path integral method. In [20] the case, where the fractional derivatives only slightly differ from the ordinary derivatives, is analyzed. Fractional Hamilton's equations are discussed in [22]. In [8] multi-order fractional differential equations are solved by using the Adomian decomposition. In several papers fractional chaotic systems are discussed. In [8] a three-dimensional fractional chaotic oscillator model is proposed. Chaotic dynamics of the fractionally damped Duffing equation is investigated in [29]. Two chaotic models for third-order chaotic nonlinear systems are analyzed in [3].

Ü. Lepik and H. Hein, *Haar Wavelets*, Mathematical Engineering,

DOI: 10.1007/978-3-319-04295-4_8,

Recently, the wavelet methods have also been used for solving problems of fractional calculus, these papers will be refereed in Sect. 8.7. The aim of this paper is to demonstrate the expediency of the Haar wavelet method for solving fractional differential and integral equations.

8.2 About the Fractional Calculus

Let us briefly consider some basic formulae about fractional calculus.

The Riemann-Liouville fractional integrals of order α are defined by

$$\left(I_{A+}^{\alpha} f\right)(x) = \frac{1}{\Gamma(\alpha)} \int_A^x f(t)(x-t)^{\alpha-1} dt, \quad (x > A,\ [\alpha] > 0) \tag{8.1}$$

and

$$\left(I_{B-}^{\alpha} f\right)(x) = \frac{1}{\Gamma(\alpha)} \int_x^B f(t)(t-x)^{\alpha-1} dt, \quad (x < B,\ [\alpha] > 0). \tag{8.2}$$

Here $\Gamma(\alpha)$ is the gamma function and $[\alpha]$ the integer part of α. The integrals (8.1–8.2) are called left-sided and right-sided fractional integrals.

As to the fractional derivatives in this paper, we shall use the Caputo derivatives defined as

$$D^{\alpha} f(x) = \frac{1}{\Gamma(n-\alpha)} \int_A^x f^{(n)}(t)(x-t)^{n-1-\alpha} dt, \tag{8.3}$$

where $f^{(n)}(x) = d^n f/dx^n$ and $n = [\alpha] + 1$.

If $\alpha + 1 = n \in N$ then $D^n f(x)$ coincides with the ordinary derivative $f^{(n)}(x)$.

The following relations hold

$$I^{\alpha}(I^{\beta} f(x)) = I^{\beta}(I^{\alpha} f(x)) = I^{\alpha+\beta} f(x), \tag{8.4}$$

$$D^{\alpha}(D^{\beta} f(x)) = D^{\beta}(D^{\alpha} f(x)) = D^{\alpha+\beta} f(x), \tag{8.5}$$

$$D^{\alpha}(I^{\beta} f(x)) = D^{\alpha-\beta} f(x),\ \alpha \ge \beta, \tag{8.6}$$

$$I^{\alpha}(D^{\alpha} f(x)) = f(x) + \sum_{k=0}^{n-1} f^{(k)}(0)\frac{x^k}{k!},\ x \ge 0. \tag{8.7}$$

8.3 Fractional Volterra Integral Equation

The fractional Volterra integral equation has the form [14]

$$u(x) - \frac{1}{\Gamma(\alpha)} \int_0^x K(x,t)(x-t)^{\alpha-1} u(t) dt = f(x), \quad 0 \leq x \leq 1. \tag{8.8}$$

The kernel $K(x,t)$ and the right-side function $f(x)$ are given, $\alpha > 0$ is a real number. The value $\alpha = 1$ corresponds to the ordinary (nonfractional) Volterra equation.

According to the Haar wavelet method, the solution of (8.8) is sought in the form $u(x) = \sum_i a_i h_i(x)$. Replacing it into (8.8) and satisfying this equation in the collocation points $x = x_l$ we obtain

$$\sum_{i=1}^{2M} a_i [h_i(x_l) - g_i(x_l)] = f(x_l), \quad l = 1, 2, ..., 2M \tag{8.9}$$

Here the symbol $g_i(x_l)$ denotes the function

$$g_i(x_l) = \frac{1}{\Gamma(\alpha)} \int_0^{x_l} K(x_l,t)(x_l-t)^{\alpha-1} h_i(t) dt \tag{8.10}$$

The matrix form of (8.9) is

$$a(H - G) = F \tag{8.11}$$

where $G(i,l) = g_i(x_l)$, $F(l) = f(x_l)$.

The solution of (8.11) is

$$a = F/(H - G) \tag{8.12}$$

The function $u(x)$ can be calculated from $u(x) = \sum_i a_i h_i(x)$.

The solution presented here is very simple. The most labor-consuming operation is the evaluation of the matrix G. In view of (2.1) the Eq. (8.11) can be rewritten in the following form (for conciseness sake, the argument i at ξ_1, ξ_2, ξ_3 is omitted):

$$\begin{aligned}
&(i) \quad G(i,l) = 0 \qquad \text{for } x_l < \xi_1, \\
&(ii) \quad G(i,l) = \frac{1}{\Gamma(\alpha)} \int_{\xi_1}^{x_l} K(x_l,t)(x_l-t)^{\alpha-1} dt \quad \text{for } x_l \in [\xi_1, \xi_2], \\
&(iii) \; G(i,l) = \frac{1}{\Gamma(\alpha)} \int_{\xi_1}^{\xi_2} K(x_l,t)(x_l-t)^{\alpha-1} dt - \frac{1}{\Gamma(\alpha)} \int_{\xi_2}^{x} K(x_l,t)(x_l-t)^{\alpha-1} dt \\
&\qquad\qquad \text{for } x_l \in [\xi_2, \xi_3], \\
&(iv) \quad G(i,l) = \frac{1}{\Gamma(\alpha)} \int_{\xi_1}^{\xi_2} K(x_l,t)(x_l-t)^{\alpha-1} dt - \frac{1}{\Gamma(\alpha)} \int_{\xi_2}^{\xi_3} K(x_l,t)(x_l-t)^{\alpha-1} dt \\
&\qquad\qquad \text{for } x_l \geq \xi_3 .
\end{aligned} \tag{8.13}$$

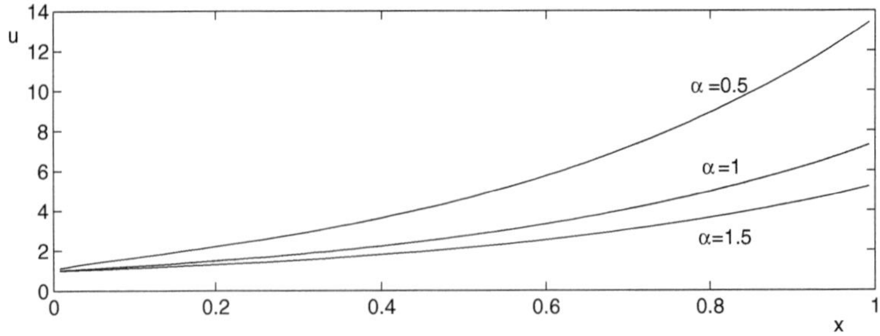

Fig. 8.1 Solution of the Volterra integral equation (8.8) for $K(x, t) = \exp(x - t)$

Table 8.1 Error estimates for the Eq. (8.8); $\alpha = 0.5$.

J	$2M$	δ_{ex}	σ_{ex}
3	16	0.332	4.5E-2
4	32	0.186	1.7E-2
5	64	0.010	6.3E-3

The integrals in (8.13) can be evaluated by some numerical techniques; but for some simplier forms of $K(x_l, t)$ analytical integration is possible.

Example 8.1: Consider the case $K(x, t) = \exp(x - t)$. Let us introduce the function

$$\varphi(x_l, \gamma_1, \gamma_2) = \int_{\gamma_1}^{\gamma_2} (x_l - t)^{\alpha-1} e^{-t} dt \tag{8.14}$$

Now (8.13) can be put into the form

$$\begin{aligned}
&(i) \quad G(i, l) = 0 && \text{for } x_l < \xi_1 \,, \\
&(ii) \quad G(i, l) = \frac{1}{\Gamma(\alpha)} \exp(x_l)\varphi(x_l, \xi_1, x_l) && \text{for } x_l \in [\xi_1, \xi_2] \,, \\
&(iii) \; G(i, l) = \frac{1}{\Gamma(\alpha)} \exp(x_l)[\varphi(x_l, \xi_1, \xi_2) - \varphi(x_l, \xi_2, x_l)] && \text{for } x_l \in [\xi_2, \xi_3] \,, \\
&(iv) \quad G(i, l) = \frac{1}{\Gamma(\alpha)} \exp(x_l)[\varphi(x_l, \xi_1, \xi_2) - \varphi(x_l, \xi_2, \xi_3)] && \text{for } x_l > \xi_3 \,.
\end{aligned} \tag{8.15}$$

These integrals were evaluated with the aid of the MatLab program QUAD.

Firstly, let us consider the nonfractional case $\alpha = 1$; here (8.8) has the exact solution $u_{ex} = exp(2x)$. Computer simulation for $J = 5$ (64 collocation points) gave the error estimates $\delta_{ex} = 8.9E-4$, $\sigma_{ex} = 5.0E-5$; for $J = 6$ (128 collocation points) we found $\delta_{ex} = 2.3E - 4$, $\sigma_{ex} = 9.0E - 6$.

Results of the calculations, which were carried out for some values of α, are plotted in Fig. 8.1. Error estimates were computed for $\alpha = 0.5$; these results are presented in Table 8.1.

8.4 Fractional Harmonic Vibrations

Consider the equation

$$D^{\alpha}u(x) + \lambda D^{\beta}u(x) + \nu u(x) = f(x), \quad x \in [0, B] \tag{8.16}$$

where $1 < \alpha < 2,\ 0 < \beta < 1,\ \lambda,\ \nu$ are prescribed constants, $f(x)$ is the forcing term. To (8.16) belong the initial conditions $u(0) = u_0,\ u'(0) = v_0$. If $\alpha = 2$ and $\beta = 1$ we get the usual differential equation of the harmonic oscillator.

The symbols $D^{\alpha},\ D^{\beta}$ denote left-sided Caputo derivatives, which are defined by (8.3). Since in the present case $n_{\alpha} = [\alpha] + 1 = 2,\ n_{\beta} = [\beta] + 1 = 1$, Eq. (8.16) gets the form

$$\frac{1}{\Gamma(2-\alpha)}\int_0^x (x-t)^{1-\alpha}u''(t)dt + \frac{\lambda}{\Gamma(1-\beta)}\int_0^x (x-t)^{-\beta}u'(t)dt + \nu u(x) = f(x) \tag{8.17}$$

This is a Volterra integral equation. Let us solve it by the Haar wavelet method.

$$\begin{aligned} u''(x) &= \sum_{i=1}^{2M} a_i h_i(x) \\ u'(x) &= \sum_{i=1}^{2M} a_i p_{1,i}(x) + u'(0) \\ u(x) &= \sum_{i=1}^{2M} a_i p_{2,i}(x) + u'(0)x + u(0)\,, \end{aligned} \tag{8.18}$$

where the functions $h_i,\ p_{1,i},\ p_{2,i}$ are calculated from (2.1), (2.6), (2.7). Replacing (8.18) into (8.17) and satisfying this equation in the collocation points x_l, we find

$$\sum_{i=1}^{2M} a_i G(i,l) = f(x_l) - \nu(v_0 x_l + u_0) - \frac{\lambda}{\Gamma(2-\beta)} v_0 x_l^{1-\beta} \tag{8.19}$$

where

$$G(i,l) = \frac{1}{\Gamma(2-\alpha)}\int_0^{x_l} (x_l - t)^{1-\alpha} h_i(t)dt + \frac{\lambda}{\Gamma(1-\beta)}\int_0^{x_l} (x_l - t)^{-\beta} p_{1,i}(t)dt + \nu p_{2,i}(x_l) \tag{8.20}$$

The integrals in (8.20) can be evaluated with the aid of (2.1), (2.6), (2.7). By doing this, we get the following formulae:

$$\begin{aligned}
G(i,l) &= 0 && \text{for } x_l < \xi_1\,,\\
G(i,l) &= \frac{1}{\Gamma(3-\alpha)}(x_l-\xi_1)^{2-\alpha} + \lambda N(\beta)(x_l-\xi_1)^{2-\beta}\\
&\quad + 0.5\nu(x_l-\xi_1)^2 && \text{for } x_l \in [\xi_1,\xi_2]\,,\\
G(i,l) &= \frac{1}{\Gamma(3-\alpha)}\left[(x_l-\xi_1)^{2-\alpha} - 2(x_l-\xi_2)^{2-\alpha}\right]\\
&\quad - \lambda N(\beta)\left[(x_l-\xi_1)^{2-\beta} - 2(x_l-\xi_2)^{2-\beta}\right]\\
&\quad + 0.5\nu[D-(\xi_3-x_l)^2] && \text{for } x_l \in [\xi_2,\xi_3]\,,\\
G(i,l) &= \frac{1}{\Gamma(3-\alpha)}\left[(x_l-\xi_1)^{2-\alpha} + (x_l-\xi_3)^{2-\alpha} - 2(x_l-\xi_2)^{2-\alpha}\right]\\
&\quad + \lambda\Big\{N(\beta)(x_l-\xi_1)^{2-\beta} - \frac{1}{\Gamma(2-\beta)}(x_l-\xi_1)(x_l-\xi_2)^{1-\beta}\\
&\quad - \frac{1}{\Gamma(2-\beta)}(\xi_3-x_l)\left[(x_l-\xi_3)^{1-\beta} - (x_l-\xi_2)^{1-\beta}\right]\\
&\quad + \frac{1}{(2-\beta)\Gamma(1-\beta)}\left[2(x_l-\xi_2)^{2-\beta}(x_l-\xi_3)^{2-\beta}\right]\Big\} + 0.5\nu D && \text{for } x_l > \xi_3\,,
\end{aligned} \tag{8.21}$$

where

$$D = (\mu\Delta x)^2, \quad N(\beta) = \frac{1}{\Gamma(2-\beta)} - \frac{1}{(2-\beta)\Gamma(1-\beta)} \tag{8.22}$$

In the following course of solution we shall again make use of the matrix representation. Now (8.19) obtains the form

$$aG = F + \Phi \tag{8.23}$$

where
$F(l) = f(l), \quad \Phi(l) = -\nu(v_0 x_l + u_0) - \frac{\lambda v_0}{\Gamma(1-\beta)} x_l^{1-\beta}$. The solution of (8.23) is

$$a = (F+\Phi)/G \tag{8.24}$$

The function u to be sought is calculated from

$$u = aP_2 + v_0 X + u_0 E \tag{8.25}$$

where $X = \{x(l)\},\ P_2 = \{p_{2,i}(x_l)\}$ and E is a $2M$ row vector of ones.

Example 8.2: Let us take for the force function $f(x) = r\sin\omega t$ and assign to parameters the values $\lambda = 0.05,\ \nu = 0.15,\ r = 1,\ \omega = 2,\ B = 30,\ u_0 = 1,\ v_0 = 0$.

In order to estimate the exactness of the presented solution, again the case $\alpha = 2$, $\beta = 1$ is considered first. The exact solution of (8.16) is

$$u_{ex}(x) = L\exp(-0.5\lambda x)\sin(\omega_1 x + \varepsilon) + R(\sin\omega x + \delta) \tag{8.26}$$

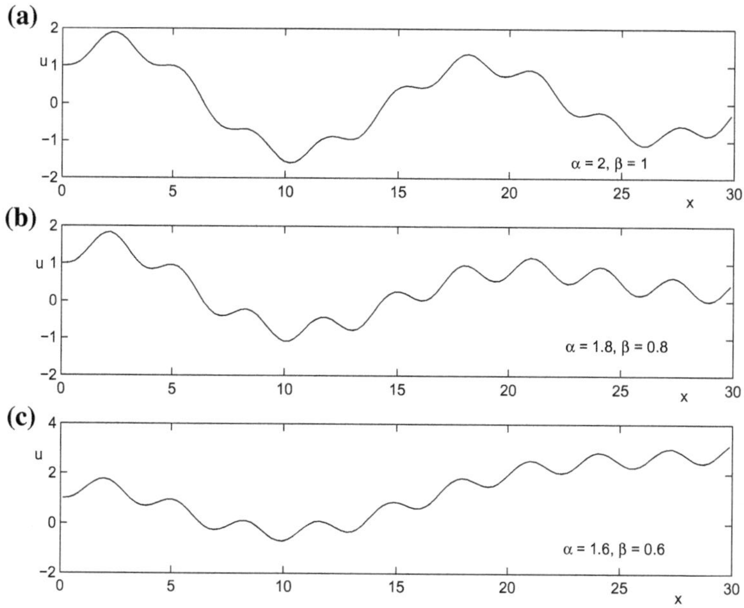

Fig. 8.2 Solution of (8.16) for different values of α and β

where

$$\omega_1 = \sqrt{\nu - 0.25\lambda^2}, \quad \tan\delta = -\frac{\lambda\omega}{\nu - \omega^2}, \\ R = \frac{r}{\sqrt{\lambda^2\omega^2 + (\nu - \omega^2)^2}} \tag{8.27}$$

The coefficients $L,\ \varepsilon$ are calculated from the boundary conditions $u_0 = 1,\ \ v_0 = 0$.

The wavelet solution was computed for $J = 6$. The error estimates gave $\delta_{ex} = 0.005$, $\sigma_{ex} = 2.4E - 4$; consequently the value $J = 6$ guarantees the necessary accuracy of the results. The function $u_{ex}(x)$ is plotted in Fig. 8.2.

Computer simulations were carried out for different values of α and β. The results are presented in Figs. 8.2, 8.3 and 8.4.

Error estimates $\delta_J,\ \sigma_J$ were calculated for the case $\alpha = 1.6,\ \beta = 0.6$. It followed from the computations that $\delta_4 = 0.789,\ \sigma_4 = 0.068,\ \delta_5 = 0.075,\ \sigma_5 = 0.005$. With a subsequent increase of the resolution level J, these estimates became worse. The reason for this lies in solving (8.23). If we calculate the determinant $\mid G \mid$, then we obtain $\mid G \mid = 1.91E + 7$ for $J = 4$, $\mid G \mid = 3.26E + 6$ for $J = 5$, $\mid G \mid = 3.6E - 3$ for $J = 6$, $\mid G \mid = 1.59E - 36$ for $J = 7$. Consequently, we have here the unfavourable case, where the matrix $\mid G \mid$ turns out to be nearly singular.

Equation (8.16) was treated in several papers from which we cite here [1, 2, 8, 21, 28, 29]. In [1, 2] the case $1 < \alpha \leq 2,\ \ \lambda = 0$ (our symbols) was analyzed, in [2] it is assumed that $f \equiv 0$; in [1] different cases of the forcing function $f(x)$ were discussed. In [29] Eq. (8.16) was solved by the Fourier transform for $\alpha = 2$,

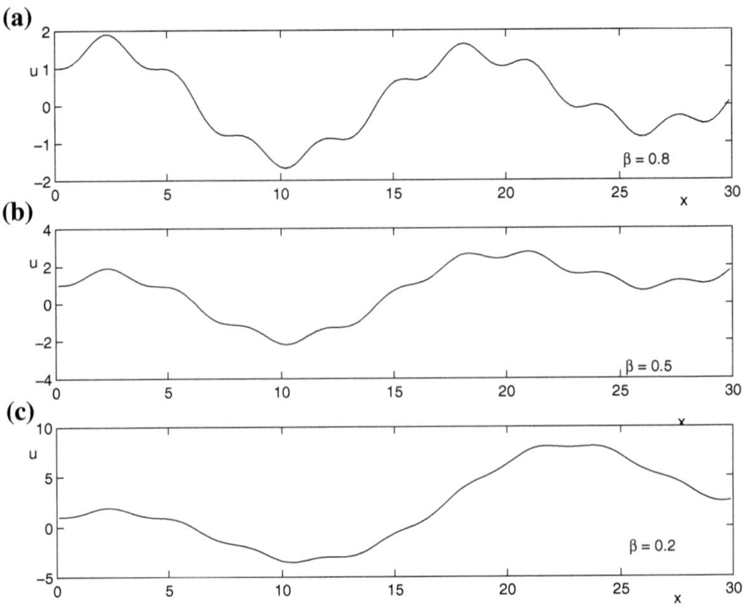

Fig. 8.3 Solution of (8.16) for $\beta = 1$

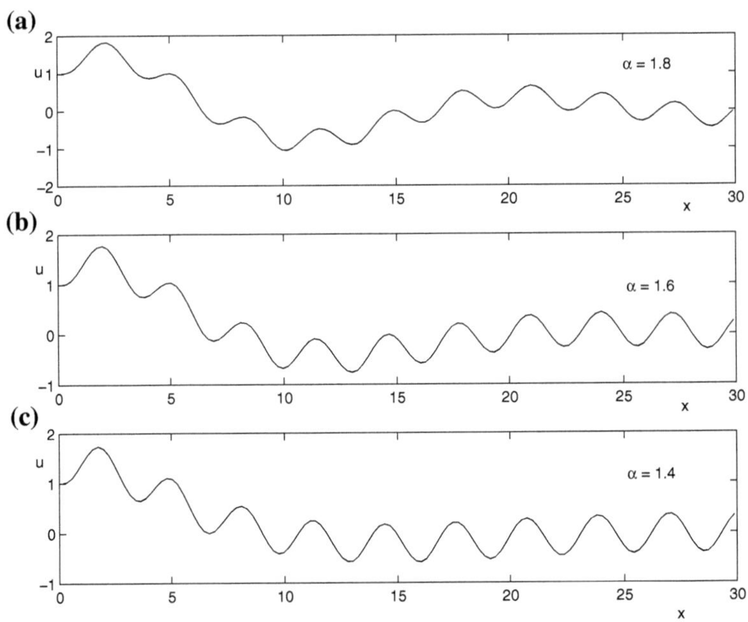

Fig. 8.4 Solution of (8.16) for $\alpha = 2$

$0 < \beta < 2, \quad \beta \neq 1$; the external force is impulsive. The almost free damping oscillator for which $\alpha = 2 + \varepsilon, \quad \beta = 1 + 0.5\varepsilon, \quad f = 0, \quad |\varepsilon| << 1$ was discussed in [21]; particular solutions of (8.16) were sought in the form $u = \exp(\lambda x)$.

8.5 Fractional Fredholm Integral Equation

The usual Fredholm integral equation has the form

$$u(x) - \int_A^B K(x,t)u(t)dt = f(x), \quad x \in [A, B] \tag{8.28}$$

If we want to get its fractional analogue, we must take into account the fact that in the case of non-integer α, the expression $(x-t)^\alpha$ has sense only for $x \geq t$. Therefore it seems reasonable to define the fractional Fredholm integral equation in the form

$$u(x) - \frac{1}{\Gamma(\alpha)}\Big[\int_A^x (x-t)^{\alpha-1}K(x,t)u(t)dt + \int_x^B (t-x)^{\alpha-1}K(x,t)u(t)dt\Big] = f(x) \tag{8.29}$$

In (8.29) stand the left-sided and right-sided Riemann-Liouville integrals, respectively. As in Sect. 8.3 the solution of (8.29) is sought in the form $u = aH$. Let us replace $x \to x_l$ and introduce the matrix

$$G(i,l) = \frac{1}{\Gamma(\alpha)}\Big[\int_A^{x_l} (x_l-t)^{\alpha-1}K(x_l,t)h_i(t)dt + \int_{x_l}^B (t-x_l)^{\alpha-1}K(x_l,t)h_i(t)dt\Big] \tag{8.30}$$

On the grounds of (2.1) this integral can be evaluated either analytically or numerically.

Now (8.29) can be put into the form

$$a(H - G) = F \tag{8.31}$$

where $H(i,l) = h_i(x_l), \quad F(l) = f(x_l)$. The solution of (8.31) is $a = F/(H - G)$.

Example 8.3: Consider the equation

$$\begin{aligned} u(x) - \frac{1}{\Gamma(\alpha)}\Big[&\int_0^x (2-x-t)(x-t)^{\alpha-1}u(t)dt \\ &+ \int_x^3 (2-x-t)(t-x)^{\alpha-1}u(t)dt\Big] = x^2 + 15/4 \end{aligned} \tag{8.32}$$

Now the kernel is $K(2 - x - t)$ and evaluating the integrals in (8.29) we find

Table 8.2 Error estimates for the Eq. (8.32); $\alpha = 1.5$

J	$2M$	δ_{ex}	σ_{ex}
3	16	0.386	4.3E-2
4	32	0.200	1.5E-2
5	64	0.100	5.4E-3
6	128	0.050	1.9E-3
7	256	0.026	6.8E-4

$$\begin{aligned}
(i)\ G(i,l) &= \frac{2(1-x_l)}{\Gamma(\alpha+1)}\big[2(\xi_2-x_l)^{\alpha}-(\xi_1-x_l)^{\alpha}-(\xi_3-x_l)^{\alpha}\big]\\
&\quad+\frac{\alpha}{\Gamma(\alpha+2)}\big[-2(\xi_2-x_l)^{\alpha+1}+(\xi_1-x_l)^{\alpha+1}+(\xi_3-x_l)^{\alpha+1}\big]\\
&\text{for } x_l<\xi_1\,,\\
(ii)\ G(i,l) &= \frac{2(1-x_l)}{\Gamma(\alpha+1)}\big[(x_l-\xi_1)^{\alpha}+2(\xi_2-x_l)^{\alpha}-(\xi_3-x_l)^{\alpha}\big]\\
&\quad+\frac{\alpha}{\Gamma(\alpha+2)}\big[(x_l-\xi_1)^{\alpha+1}-2(\xi_2-x_l)^{\alpha+1}+(\xi_3-x_l)^{\alpha+1}\big]\\
&\text{for } x_l\in[\xi_1,\xi_2]\,,\\
(iii)\ G(i,l) &= \frac{2(1-x_l)}{\Gamma(\alpha+1)}\big[(x_l-\xi_1)^{\alpha}-2(x_l-\xi_2)^{\alpha}-(\xi_3-x_l)^{\alpha}\big]\\
&\quad+\frac{\alpha}{\Gamma(\alpha+2)}\big[(x_l-\xi_1)^{\alpha+1}-2(x_l-\xi_2)^{\alpha+1}+(\xi_3-x_l)^{\alpha+1}\big]\\
&\text{for } x_l\in[\xi_2,\xi_3]\,,\\
(iv)\ G(i,l) &= \frac{2(1-x_l)}{\Gamma(\alpha+1)}\big[(x_l-\xi_1)^{\alpha}-2(x_l-\xi_2)^{\alpha}+(x_l-\xi_3)^{\alpha}\big]\\
&\quad+\frac{\alpha}{\Gamma(\alpha+2)}\big[(x_l-\xi_1)^{\alpha+1}-2(x_l-\xi_2)^{\alpha+1}+(x_l-\xi_3)^{\alpha+1}\big]\\
&\text{for } x_l\geq\xi_3\,.
\end{aligned} \tag{8.33}$$

Let us begin from the case $\alpha = 1$, which corresponds to the non-fractional equation. This equation has the exact solution

$$u_{ex}(x) = x^2 - 3, \quad x \in [0, 3] \tag{8.34}$$

The computer simulation for $\alpha = 1$ gave the following error estimates

$$\begin{aligned}
(i)\ \delta_{ex} &= 7.7E-3, \quad \sigma_{ex} = 1.8E-4 \ \text{for} J = 5\,,\\
(ii)\ \delta_{ex} &= 6.8E-4, \quad \sigma_{ex} = 3.2E-5 \ \text{for} J = 6\,.
\end{aligned}$$

The results of the computations for $\alpha = 0.5$, $\alpha = 0.9$, $\alpha = 1.0$, $\alpha = 1.1$, $\alpha = 1.5$ are plotted in Fig. 8.5.

Error analysis was carried out for $\alpha = 1.5$, the results are given in Table 8.2.

Calculations show that for this case the determinant $\mid H - G \mid$ essentially differs from zero (e.g. $\mid H - G \mid = -9.7E + 18$ for $J = 4$ and $\mid H - G \mid = -5.6E + 153$, consequently, no problems for solving (8.31) occur.

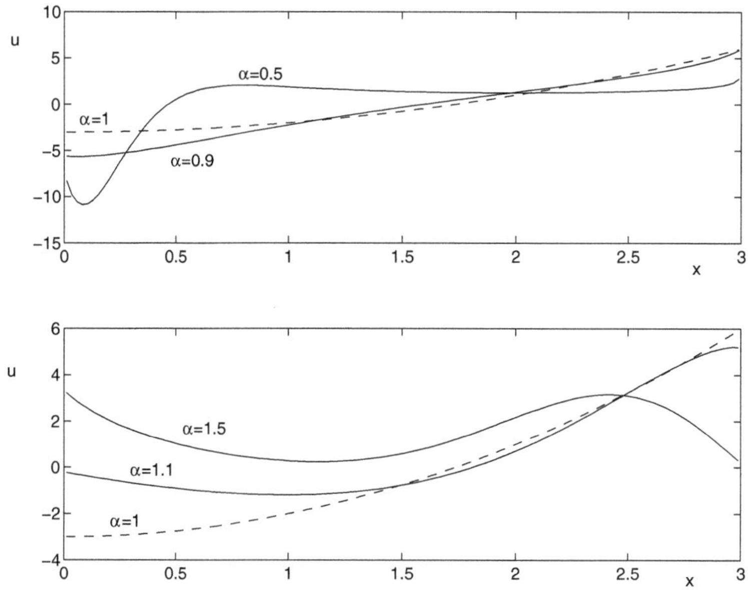

Fig. 8.5 Solution of the Fredholm equation (8.32) for different values of α

8.6 Haar Wavelet Operational Method for Solving Fractional ODEs

An alternative method for solving fractional ODEs exists and is based on the *Haar wavelet operational matrix of fractional order integration* [11]. The following presentation is based on the paper by Li and Zhao [19]. To make the system of notations uniform with this book, some symbols of the Li and Zhao paper are changed.

To begin with we define the solution vector

$$h(x) = [h_1(x), h_2(x), \ldots, h_{2M}(x)]^T. \tag{8.35}$$

Here $h_i(x)$ denotes the i-th Haar wavelet defined by (2.1). Calculating $h_i(x)$ in the collocation points x_l we obtain the Haar matrix with the elements $H(i, l) = h_i(x_l)$.

The fractional order integral of Haar functions can be written in the form

$$I^\alpha h(x) = P^\alpha h(x), \tag{8.36}$$

where P^α is a $2M \times 2M$ matrix. It is named the *Haar wavelet operational matrix of integration*.

For calculating this matrix we make use of the block pulse functions (BPF):

$$b_i(x) = \begin{cases} 1 \text{ for } (i-1)\Delta x \le x \le i\Delta x\,, \\ 0 \text{ elsewhere}\,, \end{cases} \tag{8.37}$$

Here $\Delta x = 1/(2M),\ 1 = 1, 2, \ldots, 2M - 1$.

Let us define the vector

$$B(x) = [b_1(x), b_2(x), \ldots, b_{2M-1}(x)]^T. \tag{8.38}$$

The Haar functions can be expanded into the series of block pulse functions as

$$h(x) = HB(x). \tag{8.39}$$

Here H denotes the Haar matrix with the components $H(i,l) = h_i(x_l)$, where $x_l,\ l = 1, 2, \ldots, 2M$ are the collocation points.

Kilicman and Zhour [15] defined the *block pulse operation matrix of fractional order integration* F^α as following

$$I^\alpha B(x) = F^\alpha B(x). \tag{8.40}$$

They demonstrated that this matrix can be calculated according to the formulae

$$F^\alpha = \frac{1}{M^\alpha}\frac{1}{\Gamma(\alpha+2)}\begin{pmatrix} 1 & \eta_1 & \eta_2 & \ldots & \eta_{2M-1} \\ 0 & 1 & \eta_1 & \ldots & \eta_{2M-2} \\ 0 & 0 & 1 & \ldots & \eta_{2M-3} \\ & & & \ldots & \\ 0 & 0 & 0 & \ldots & 1 \end{pmatrix}, \tag{8.41}$$

where $\eta_\nu = (\nu+1)^{\alpha+1} - 2\nu^{\alpha+1} + (\nu-1)^{\alpha+1},\ \nu = 1, 2, \ldots, 2M-1$ and $\Gamma(\alpha)$ is the Gamma function.

In view of (8.39) and (8.40) we have

$$I^\alpha h(x) = P^\alpha h(x) = P^\alpha HB(x). \tag{8.42}$$

On the other hand

$$I^\alpha h(x) = I^\alpha HB(x) = H(I^\alpha B(x)) = HF^\alpha B(x) \tag{8.43}$$

and therefore

$$P^\alpha H = HF^\alpha. \tag{8.44}$$

Consequently

$$P^\alpha = HF^\alpha H^{-1}. \tag{8.45}$$

If we have to solve a fractional ODE, then we shall expand the highest derivative in this equation into the Haar series

$$D^{(\alpha)}y(x) = ah(x). \tag{8.46}$$

Here a denotes the Haar wavelet coefficients vector

$$a = [a_1, a_2, \ldots, a_{2M}]. \tag{8.47}$$

Making use of (8.4)–(8.7) we calculate the lower derivatives and the function $y(x)$. All these results are substituted into the differential equation. In this way we get an algebraic system of equations for evaluating the Haar wavelet coefficients a_i. Consider some examples.

Example 8.4: Solve the Bagley-Torvik equation

$$D^2y(x) + D^{1.5}y(x) + y(x) = 1 + x,\ x \in [0, 1], \tag{8.48}$$

subject to initial conditions $y(0) = 1$, $y'(0) = 1$. This equation has an exact solution $y(x) = x + 1$.

Let us take

$$D^2y(x) = ah(x). \tag{8.49}$$

In view of (8.4)–(8.7) we have

$$\begin{aligned} &D^{1.5}y(x) = (I^{0.5}D^2y)x = aP^{0.5}h(x) = aHF^{0.5}H^{-1}h(x), \\ &Dy(x) = aP^1h(x) + y'(0) = aP^1h(x) + 1, \\ &y(x) = aP^2h(x) + [1, 1, \ldots, 1]H^{-1}P^1h(x) + 1. \end{aligned} \tag{8.50}$$

Next we expand the function $f(x) = 1 + x$ by the Haar functions $f(x) = ch(x)$, where c is a constant vector. In view of (2.1) we find

$$c_i = m\int_0^1 f(x)h_i(x)dx = m\int_0^1 (1 + x)h_i(x)dx,\ m = 2^j. \tag{8.51}$$

Substituting all these results into (8.41) we get

$$ah(x)+aP^{0.5}h(x)+aP^2h(x)+[1, 1, \ldots, 1]H^{-1}P^1h(x)+[1, 1, \ldots, 1] = ch(x). \tag{8.52}$$

Satisfying this equation in the collocation points we obtain an algebraic system for calculating the coefficients a.

It is mentioned by Li and Zhao [19] that this result is in perfect agreement with the exact solution.

Example 8.5: Consider the equation

$$D^{1.8}y(x) + 0.05Dy(x) + 0.15y(x) = \sin(2x), \tag{8.53}$$

subject to $y(0) = 1$, $y'(0) = 0$. This is a special case from the paper by Lepik [16] and it is discussed also in Sect. 8.4 of the present book.

Let us take

$$D^{1.8}y(x) = ah(x). \tag{8.54}$$

Now we have

$$\begin{aligned} Dy(x) &= (I^{0.8}D^{1.8}y)(x) = aP^{0.8}h(x), \\ y(x) &= aP^{1.8}h(x) + y(0). \end{aligned} \tag{8.55}$$

The function $f(x) = \sin(2x)$ is expanded into the Haar series $f(x) = ch(x)$, where the constant vector c is calculated as shown in the case of Example 8.4.

Substituting all these results into (8.46) we get

$$ah(x) + 0.05aP^{0.8}h(x) + 0.15aP^{1.8}h(x) + 0.15 = ch(x). \tag{8.56}$$

Again this result is transformed into a system of algebraic equations from which the vector a is calculated. Li and Zhao [19] note that the numerical results are in very good agreement with our results [16].

Now let us compare both methods. In solving fractional ODEs, we have reduced the differential equation to a fractional Volterra integral equation (8.8). In the case of the operational matrix method, the main problem is to calculate the operational matrix of the fractional order integration. Both methods give results which are in good accordance with the existing solutions but the operational matrix method seems to be mathematically more simple.

Otherwise, in the case of fractional integral equations, the operational matrix method is not directly applicable and we do not know any paper in which this approach is used for getting Haar wavelet solutions.

8.7 Related Papers

There are not many papers in which the wavelet methods are applied for solution of problems of fractional calculus. We start our overview with papers in which the Haar wavelet was applied for solving fractional ODEs and PDEs. Besides the paper by Li and Zhao [19] which was referred to in Sect. 8.6, we would note here the paper by Rehman and Khan [25] in which boundary value problems of fractional ODEs are investigated. Ray [23] considered the Bagley-Torvik equation, which models the motion of an immersing plate in a Newtonian fluid. Error analysis for numerical solution of an ODE with variable coefficient

$$D^{\alpha}u(t) + \alpha(t)u(t) = f(x) \tag{8.57}$$

was presented by Chen et al. [7].

An algorithm for calculating fractional order integrals was given by Gao and Liao [10]. Saeedi et al. [27] presented a method based on Haar wavelets for solving fractional Volterra integral equations.

In some papers [26, 32] the Haar method is used for solving the fractional PDEs. Convection-diffusion equations with variable coefficients

$$\frac{\partial^{\alpha} u(x,t)}{\partial t^{\alpha}} = -b(x)\frac{\partial u(x,t)}{\partial x} + \alpha(x)\frac{\partial^{\beta} u(x,t)}{\partial x^{\beta}} + \eta(x,t) \tag{8.58}$$

were solved by Chen et al. [6].

In all these papers, the operational matrix method is applied and the Haar wavelet operational matrix of the fractional order integration P^{α} is put together.

Next, the solutions based on other wavelet families are discussed. Evidently, one of the first papers in this field belongs to Al-Assaf et al. [4], in which fractional chaotic system parameters were identified making use of Daubechies wavelets.

Shannon wavelets for solving problems of fractional calculus were taken into use by Cattani [5]. An operational matrix of generalized hat functions was put together by Tripathi et al. [30]. An analytical method based on Walsh functions was applied for solving the fractional transport equation by Kadem and Baleanu [13]. Generalized block pulse functions were used for solving fractional ODEs by Li and Sun [18]. Solutions based on the Legendre wavelets were presented by Jafari et al. [12] and also by Rehman and Khan [24].

Solutions of fractional ODEs by the second kind of Chebyshev wavelets were obtained by Wang and Fan [31]. The Cubic B-spline wavelet collocation method was used by Li [17].

References

1. Achar, B., Hanneken, J., Clarke, T.: Response characteristics of a fractional oscillator. Phys. A Stat. Mech. Appl. **309**, 275–288 (2002)
2. Achar, B., Hanneken, J., Enck, T., Clarke, T.: Dynamics of the fractional oscillator. Phys. A Stat. Mech. Appl. **297**, 361–367 (2001)
3. Ahmad, W., Sprott, J.: Chaos in fractional-order autonomous nonlinear systems. Chaos, Solitons Fractals **16**(2), 339–351 (2003)
4. Al-Assaf, Y., El-Khazali, R., Ahmad, W.: Identification of fractional chaotic system parameters. Chaos, Solitons Fractals **22**(4), 897–905 (2004)
5. Cattani, C.: Fractional calculus and Shannon wavelet. Math. Probl. Eng. **2012** (2012). doi:10.1155/2012/502812
6. Chen, Y., Wu, Y., Cui, Y., Wang, Z., Jin, D.: Wavelet method for a class of fractional convection-diffusion equation with variable coefficients. J. Comput. Sci. **1**(3), 146–149 (2010)
7. Chen, Y., Yi, M., Yu, C.: Error analysis for numerical solution of fractional differential equation by Haar wavelets method. J. Comput. Sci. **3**(5), 367–373 (2012)
8. Daftardar-Gejji, V., Jafari, H.: Solving a multi-order fractional differential equation using Adomian decomposition. Appl. Math. Comput. **189**(1), 541–548 (2007)
9. Eab, C., Lim, S.: Path integral representation of fractional harmonic oscillator. Phys. A Stat. Mech. Appl. **371**(2), 303–316 (2006)
10. Gao, Z., Liao, X.: Discretization algorithm for fractional order integral by Haar wavelet approximation. Appl. Math. Comput. **218**(5), 1917–1926 (2011)
11. Guf, J., Jiang, W.: The Haar wavelets operational matrix of integration. Int. J. Syst. Sci. **27**(7), 623–628 (1996)

12. Jafari, H., Yousefi, S., Firoozjaee, M., Momani, S., Khalique, C.: Application of Legendre wavelets for solving fractional differential equations. Comput. Math. Appl. **62**(3), 1038–1045 (2011)
13. Kadem, A., Baleanu, D.: Analytical method based on Walsh function combined with orthogonal polynomial for fractional transport equation. Commun. Nonlinear Sci. Numer. Simul. **15**, 491–501 (2010)
14. Kilbas, A., Srivastava, H., Trujillo, J.: Theory and Applications of Fractional Differential Equations. Elsevier Science Ltd, Amsterdam (2006)
15. Kilicman, A., Zhour, Z.: Kronecker operational matrices for fractional calculus and some applications. Appl. Math. Comput. **187**(1), 250–265 (2007)
16. Lepik, Ü.: Solving fractional integral equations by the Haar wavelet method. Appl. Math. Comput. **214**(2), 468–478 (2009)
17. Li, X.: Numerical solution of fractional differential equations using cubic B-spline wavelet collocation method. Commun. Nonlinear Sci. Numer. Simul. **17**(10), 3934–3946 (2012)
18. Li, Y., Sun, N.: Numerical solution of fractional differential equations using the generalized block pulse operational matrix. Comput. Math. Appl. **62**, 1046–1054 (2011)
19. Li, Y., Zhao, W.: Haar wavelet operational matrix of fractional order integration and its applications in solving the fractional order differential equations. Appl. Math. Comput. **216**(8), 2276–2285 (2010)
20. Miyakoda, T.: On an almost free damping vibration equation using N-fractional calculus. J. Comput. Appl. Math. **144**(12), 233–240 (2002)
21. Oldham, K., Spanier, J.: The Fractional Calculus. Academic Press, New York (1974)
22. Rabei, E., Nawafleh, K., Hijjawi, R., Muslih, S., Baleanu, D.: The Hamilton formalism with fractional derivatives. J. Math. Anal. Appl. **327**(2), 891–897 (2007)
23. Ray, S.: On Haar wavelet operational matrix of general order and its application for the numerical solution of fractional Bagley-Torvik equation. Appl. Math. Comput. **218**(9), 5239–5248 (2012)
24. Rehman, M., Khan, R.: The Legendre wavelet method for solving fractional differential equations. Commun. Nonlinear Sci. Numer. Simul. **16**(11), 4163–4173 (2011)
25. Rehman, M., Khan, R.: A numerical method for solving boundary value problems for fractional differential equations. Appl. Math. Model. **36**(3), 894–907 (2012)
26. Rehman, M., Khan, R.: Numerical solutions to initial and boundary value problems for linear fractional partial differential equations. Appl. Math. Model. **37**(7), 5233–5244 (2013)
27. Saeedi, H., Mollahasani, N., Moghadam, M., Chuev, G.: An operational Haar wavelet method for solving fractional Volterra integral equations. Int. J. Appl. Math. Comput. Sci. **21**, 535–547 (2011)
28. Schäfer, I., Kempfle, S.: Impulse responses of fractional damped systems. Nonlinear Dyn. **38**, 61–68 (2004)
29. Sheu, L., Chen, H., Chen, J., Tam, L.: Chaotic dynamics of the fractionally damped Duffing equation. Chaos, Solitons Fractals **32**(4), 1459–1468 (2007)
30. Tripathi, M., Baranwal, V., Pandey, R., Singh, O.: A new numerical algorithm to solve fractional differential equations based on operational matrix of generalized hat functions. Commun. Nonlinear Sci. Numer. Simul. **18**, 1327–1340 (2013)
31. Wang, Y., Fan, Q.: The second kind Chebyshev wavelet method for solving fractional differential equations. Appl. Math. Comput. **218**(17), 8592–8601 (2012)
32. Wu, J.: A wavelet operational method for solving fractional partial differential equations numerically. Appl. Math. Comput. **214**, 31–40 (2009)

Chapter 9
Applying Haar Wavelets in the Optimal Control Theory

9.1 Basic Elements of Optimal Control

In this chapter, the Haar wavelet method for solving problems of constrained optimal control is applied. To begin with, we are reminded of some of the necessary conditions for the optimal control.

Performance index. We have to find the control history $u(t)$ that minimizes the functional

$$I = \int_{t_0}^{T} F(t, x, u)dt . \tag{9.1}$$

Here t, t_0, T are scalars, the state variables $x(t)$ and controls $u(t)$ are n- and r-dimensional vectors, respectively. It is assumed that x, u have as many derivatives as are needed for the theory being developed.

State variables. They are subjected to differential constraints

$$\dot{x} = f(t, x, u) . \tag{9.2}$$

Adjoint variables. If $\hat{H}$ denotes the Hamiltonian,

$$\hat{H}(t, x, u, \psi) = -F + \psi f , \tag{9.3}$$

then the adjoint variables (Lagrange multipliers) are calculated from the equations

$$\dot{\psi} = -\frac{\partial \hat{H}}{\partial x} . \tag{9.4}$$

Here f and ψ are n-dimensional vectors.

Ü. Lepik and H. Hein, *Haar Wavelets*, Mathematical Engineering,
DOI: 10.1007/978-3-319-04295-4_9,

Admissible controls. If the vector u is unconstrained then optimal controls are calculated from

$$\frac{\partial \hat{H}}{\partial u} = 0 . \tag{9.5}$$

If $u \in U$, where U is a closed set, the maximum principle (Pontryagin's principle) holds

$$\hat{H}(t,\ x(t),\ u(t),\ \psi(t)) = \max_{u \in U} \hat{H}(t,\ x(t),\ u,\ \psi(t)) . \tag{9.6}$$

For simplicity of discussion, in following we consider only the case where u is a scalar satisfying the inequality constraint $|u(t)| < u_0$.

Boundary conditions. We confine ourselves with the case where the values $x_\alpha(t_0)$, $\alpha = 1, 2, ..., \nu$ are prescribed and the remained values $x_\alpha(t_0)$, $\beta = 1, 2, ..., n - \nu$ are free. It follows from the transversality conditions that $\psi_\alpha(t_0)$ is free and $\psi_\beta(t_0) = 0$. Similar results hold also for the final time $t = T$.

Free final time. If the final time T is not fixed the complementary condition $\hat{H}|_{t=T} = 0$ holds. If $F = 1$ we get a minimum time problem $T = \min$.

Integral constraint. A scalar integral constraint has the form

$$\int_{t_0}^{T} G(\,x,\ u)dt = K\,. \tag{9.7}$$

If this case, it is suitable to introduce a new state variable

$$\dot{x}_{n+1} = G(\,x,\ u\,)\,, \quad x_{n+1}(t_0) = 0\,, \quad x_{n+1}(t) = K\,. \tag{9.8}$$

State inequality constraint. Consider the case

$$g(t,\ x) \le 0\,, \tag{9.9}$$

where g is a prescribed function. It is assumed that the equality $g(t,\ x) = 0$ holds for $t \in [t_1, t_2]$. In this case, the adjoint system gets the form

$$\dot{\psi} = -\frac{\partial H}{\partial x} + \mu \nabla S\,, \quad t \in [t_1, t_2]\,, \tag{9.10}$$

where

$$S(\,x,\ u) = \frac{\partial g}{\partial x} f\,, \quad \nabla S = \frac{\partial S}{\partial x}\,. \tag{9.11}$$

Here $\mu(t)$ denotes the Lagrange multiplier.

It can be shown that the following conditions hold

$$\begin{aligned} \psi(t_1+0) &= \psi(t_1-0)\,, \\ \psi(t_2+0) &= \psi(t_2-0) - \mu(t_2)\,\nabla S(\,u(t_2),\ x(t_2))\,. \end{aligned} \tag{9.12}$$

Consequently, the adjoint variables are continuous going on the boundary $g = 0$ and a jump can occur coming off the boundary.

Different generalizations of these equations are possible, for them consult a textbook about optimal control (e.g. [1, 10, 14, 15]).

The optimal control theory has many applications in contemporary engineering systems, as in structural mechanics, space flights, chemical engineering. The optimality conditions often lead to problems which are mathematically difficult to solve. To overcome these difficulties various numerical methods, such as shooting techniques, nonlinear programming, quasi-linearization, etc. have been worked out.

9.2 Method of Solution

The purpose of this chapter is to demonstrate the efficiency of the Haar wavelets for solving problems of optimal control. The main problem is to integrate the state equations (9.2) and the equations for the adjoint variables (9.4) for the corresponding boundary conditions. This is carried out by the Haar wavelet method. In several papers, the control force $u(t)$ is also expanded into the Haar series. We make use of an alternative method according to which the function $u(t)$ is calculated directly from the extremality conditions. This essentially simplifies the solution. In Sects. 9.3–9.5, three problems with different constraints are solved.

The selection of these problems is mainly conditioned from the methodological considerations: we have chosen simple problems for which mostly exact solutions are known. This allows us to estimate the exactness of the achieved results. The proposed method of solution is, of course, applicable for more complicated problems. The results of this chapter are published in [16].

9.3 Optimal Control with an Integral Constraint

Consider the problem

$$\begin{aligned} &\int_0^2 x\,dt \to \inf\,, \quad \int_0^2 \dddot{x}^2 dt = 1\,, \\ &x(0) = \dot{x}(0) = 0\,, \quad \ddot{x}(2) = 0\,. \end{aligned} \tag{9.13}$$

We interprete the acceleration $\ddot{x}$ as a control and introduce the state variables

$$x_1 = x\,, \quad x_2 = \dot{x}\,, \quad x_3 = \ddot{x} = u\,, \quad x_4 = \int_0^t u^2 dt\,.$$

The state equations are

$$\dot{x}_1 = x_2\,, \quad \dot{x}_2 = x_3\,, \quad \dot{x}_3 = u\,, \quad \dot{x}_4 = u^2 \tag{9.14}$$

with the boundary conditions $x_1(0) = x_2(0) = x_3(2) = x_4(0) = 0$, $x_4(2) = 1$.
The Hamiltonian is

$$\hat{H} = -x_1 + \psi_1\, x_2 + \psi_2\, x_3 + \psi_3\, u + \psi_4\, u^2\,. \tag{9.15}$$

The adjoint system has the form

$$\dot{\psi}_1 = E\,, \quad \dot{\psi}_2 = -\psi_1\,, \quad \dot{\psi}_3 = -\psi_2\,, \quad \dot{\psi}_4 = 0\,. \tag{9.16}$$

In this system belong the transversality conditions $\psi_1(2) = \psi_2(2) = \psi_3(2) = 0$. To begin with, we integrate (9.16) by assuming

$$\begin{aligned} &\dot{\psi}_1 = b_1\, H\,, \quad \psi_1 = b_1\, R\,, \\ &\dot{\psi}_2 = b_2\, H\,, \quad \psi_2 = b_2\, R\,, \\ &\dot{\psi}_3 = b_3\, H\,, \quad \psi_3 = b_3\, R\,, \quad \psi_4 = \lambda = \text{const}\,. \end{aligned} \tag{9.17}$$

Here H is the Haar matrix and R denotes the matrix with coefficients

$$R(i,l) = P_1(i,l) - p_{1,i}(T). \tag{9.18}$$

In view of (2.6), (2.7) we have $p_{1,i}(T) = T - t_0$ for $i = 1$ and $p_{1,i}(T) = 0$ for $i \neq 1$. Substituting (9.17) into (9.16) we obtain

$$\begin{aligned} &b_1 H = E\,, \\ &b_1 R + b_2 H = 0\,, \\ &b_2 R + b_3 H = 0\,. \end{aligned} \tag{9.19}$$

From here we calculate the wavelet coefficients b_1, b_2, b_3 and find from (9.17) the adjoint variables ψ_1, ψ_2, ψ_3.

Evaluating the optimal control from (9.5) we find

$$u = -\frac{\psi_3}{2\psi_4} = -\frac{\psi_3}{2\lambda}\,. \tag{9.20}$$

Since ψ_3 is already known, we can calculate the auxiliary variable

$$\tilde{u} = \lambda \;\; u = -\psi_3/2\,. \tag{9.21}$$

Next, the state variables are developed into the Haar series:

$$\begin{aligned} \dot{x}_1 &= a_1 H\,, & x_1 &= a_1 P_1\,, \\ \dot{x}_2 &= a_2 H\,, & x_2 &= a_2 P_1\,, \\ \dot{x}_3 &= a_3 H\,, & x_3 &= a_3 P_1\,, \\ \dot{x}_4 &= a_4 H\,, & x_4 &= a_4 P_1\,. \end{aligned} \tag{9.22}$$

Replacing these results into (9.14) we get

$$\begin{aligned} a_1 H - a_2 P_1 &= 0\,, \\ a_2 H - a_3 P_1 &= 0\,, \\ a_3 H &= \tilde{u}/\lambda\,, \\ a_4 H &= \tilde{u}^2/\lambda^2\,. \end{aligned} \tag{9.23}$$

The Lagrange multiplier λ is calculated in the following way. From $(9.23)_4$ we find

$$\tilde{a}_4 = \lambda^2\, a_4 = \tilde{u}^2/\, H\,.$$

Satisfying the boundary condition $x_4(2) = 1$ we get from (9.22)

$$x_4|_{t=2} = a_4 P_1|_{t=2} = 2a_4(1) = 1\,.$$

From these two equations we find

$$\lambda = \sqrt{2\tilde{a}_4(1)}\,. \tag{9.24}$$

To end up with the solution, the wavelet coefficients a_i and the state variables x_1, x_2 are calculated from (9.22) to (9.23).

The exact solution is

$$\begin{aligned} x_{ex} &= -\frac{1}{30\,\lambda}\frac{1}{48}(t-2)^6 - 5t^2 + 4t - 8, \\ u_{ex} &= -\frac{1}{12\lambda}(t-2)^3\,, \quad \lambda_{ex} = 0.3563. \end{aligned} \tag{9.25}$$

For estimating the accuracy of our results, the error estimates

$$\begin{aligned} \delta x &= \max_i |x_{ex}(t_i) - x(t_i)|\,, \\ \delta u &= \max_i |u_{ex}(t_i) - u(t_i)|\,, \\ \delta\lambda &= |\lambda_{ex} - \lambda| \end{aligned} \tag{9.26}$$

are introduced. Results of the computer simulation are presented in Table 9.1.

It follows from this table that already a small number of points ($J = 4$ with 32 grid points) guarantees sufficient accuracy.

Table 9.1 Error estimates for the problem (9.13)

J	δx	δu	$\delta\lambda$
4	6.5E-4	6.3E-4	3.8E-7
5	1.6E-4	1.6E-4	3.8E-7
6	4.0E-5	4.0E-5	3.8E-7

9.4 Optimal Control with a State Inequality Constraint

Bryson and Ho [1] discussed the problem

$$\begin{aligned} &I = \tfrac{1}{2}\int_0^1 u^2 dt \to \min\,, \quad x_1 \le l \\ &\dot{x}_1 = x_2\,, \quad \dot{x}_2 = u\,, \quad x_1(0) = x_1(1) = 0\,, x_2(0) = -x_2(1) = 1\,, \end{aligned} \tag{9.27}$$

where $l > 0$ is a given constant.

Consider the case where the equality $x_1(t) = l$ holds in the subinterval $t \in [t_1, t_2]$. This subinterval cannot be near the boundaries $t = 0$ or $t = 1$, since in this case the prescribed boundary conditions cannot be satisfied.

Due to symmetry we can solve the problem (9.27) only for the half-interval $t \in [0, 0.5]$. It is reasonable to assume that $x_1(t) < l$ for some interval $t \in [0, t_1]$ and $x_1(t) = l$ for $t \in [t_1, 0.5]$.

The Hamiltonian is

$$\hat{H} = -\frac{1}{2}u^2 + \psi_1 x_2 + \psi_2 u\,. \tag{9.28}$$

It follows from the optimality condition (9.5) that $\psi_2 = u$.

Next we put together the adjoint system (9.10). Since in the present case $g = x_1 - l$, $S = x_2$, $\nabla S = (0, 1)$ we find

$$\dot{\psi}_1 = 0\,, \quad \dot{\psi}_2 = -\psi_1 + \mu\,, \tag{9.29}$$

whereas $\mu(t) = 0$ for $t \in [0, t_1]$.

To begin with we consider the subinterval $t \in [t_1, 0.5]$. Since $x_1 = l$ it follows from the state equations that $x_2 = u = 0$ and, consequently, $\psi_2 = 0$. Integrating (9.29) we find $\psi_1 = C_1$, $\mu(t) = C_1$, where C_1 denotes the integration constant.

Now the problem (9.27) can be paraphrased as

$$\begin{aligned} &I = \frac{1}{2}\int_0^{t_1} u^2 dt \to \min\,, \\ &\dot{x}_1 = x_2\,, \quad \dot{x}_2 = u\,, \quad x_1(0) = 0\,, \quad x_2(0) = 1\,, \\ &x_1(t_1) = l\,, \quad x_2(t_1) = 0\,. \end{aligned} \tag{9.30}$$

The wavelet solution is sought in the matrix form

$$\begin{aligned}\dot{x}_1 &= a_1 H\,, & x_1 &= a_1 P_1\,,\\ \dot{x}_2 &= a_2 H\,, & x_2 &= a_2 P_1 + E\,,\\ \dot{\psi}_1 &= b_1 H\,, & \psi_1 &= b_1 P_1 + C_1 E\,,\\ \dot{\psi}_2 &= b_2 H\,, & \psi_2 &= b_2 P_1 + C_2 E\,.\end{aligned} \tag{9.31}$$

The matrices H and P_1 can be calculated according to (2.1), (2.6), (2.7) assuming that $A = 0$, $B = t_1$.

Replacing (9.31) into the state Eq. (9.30) and into the adjoint system (9.29) we obtain

$$\begin{aligned}a_1 H &= a_2 P_1 + E\,,\\ a_2 H &= b_2 P_1 + C_2 E\,,\\ b_1 H &= 0\,,\\ b_2 H &= -b_1 P_1 - C_1 E\,.\end{aligned} \tag{9.32}$$

Since $E/H = E_1 = [1, 0, 0, \ldots, 0]$, it follows from the third and fourth equation that $b_1 \equiv 0$, $b_2 = -C_1 E/H = -C_1 E_1$, $C_2 = C_1 t_1$. Due to the continuity of $\psi_2(t)$ we have $\psi_2(t_1) = 0$ and we obtain

$$\psi_2 = C_1(t_1 E - t) = u\,. \tag{9.33}$$

Here t is a $2M$-dimensional vector. Integration of (9.31) gives

$$\begin{aligned}a_2 H &= -C_1 E_1 P_1 + C_2 E = -C_1 t + C_2 E\,,\\ x_2 &= -C_1(t/H)\,P_1 + (E/H)P_1 + E\,.\end{aligned} \tag{9.34}$$

It can be verified by computer simulation that the following formulae hold

$$\begin{aligned}(E/H)P_1 &= E_1 P_1 = t,\\ \tfrac{1}{\alpha!}(t^\alpha/H)P_1 &= \tfrac{1}{(\alpha+1)!}t^{\alpha+1}\end{aligned} \tag{9.35}$$

In view of (9.35), Eq. $(9.34)_2$ can be put into the form

$$x_2 = C_1 t(t_1 - t/2) + 1\,. \tag{9.36}$$

Since

$$a_2 P_1 = -C_1(t/H)P_1 + C_1 t_1(E/H)P_1 = -C_1 t^2/2 + C_1 t_1 t\,,$$

we have

$$a_1 = (a_2\, P_1 + E)/\,H = -\frac{1}{2}C_1\, t^2/\,H + C_1 t_1\, t/\,H + E/\,H$$

and

$$x_1 = -\frac{1}{2} C_1 (t^2 / \ H) \ P_1 + C_1 t_1 (t / \ H) \ P_1 + (\ E / \ H) \ P_1 \, .$$

On the grounds of (9.35), this result can be rewritten in the form

$$x_1 = -C_1 t^2 \left(\frac{t}{6} - \frac{1}{2} t_1 \right) + t \, . \tag{9.37}$$

The constants C_1, t_* are calculated from the boundary conditions $x_1(t_*) = l$, $x_2(t_*) = 0$. Satisfying this conditions we find $t_1 = 3l$, $C_1 = -2/t_1^2$ and

$$\begin{aligned} x_1 &= l\tau(\tau^2 - 3\tau + 3) \, , \\ x_2 &= (1 - \tau)^2 \, , \\ u &= -\frac{2}{3l}(1 - \tau), \ \tau = t/t_1 \, . \end{aligned} \tag{9.38}$$

The same results were obtained by Bryson and Ho [1].

So we can state that the Haar wavelet method enabled us to find the exact analytical solution of the problem.

9.5 Optimal Control with a Control Inequality Constraint

Let us solve the problem

$$\begin{aligned} &I = \int_0^1 (x_1^2 + x_2^2 + \alpha u^2) dt \to \min \, , \quad |u| \le u_0 \, , \\ &\dot{x}_1 = x_2 \, , \quad \dot{x}_2 = -x_2 + u \, , \quad x_1(0) = 0 \, , \quad x_2(0) = -1 \, . \end{aligned} \tag{9.39}$$

It is assumed that the control is smooth and thus the function $u(t)$ must be continuous.

Introducing the Hamiltonian

$$\hat{H} = -(x_1^2 + x_2^2 + \alpha u^2) + \psi_1 x_2 + \psi_2(-x_2 + u) \tag{9.40}$$

we put together the adjoint system

$$\dot{\psi}_1 = -\frac{\partial \hat{H}}{\partial x_1} = 2x_1 \, , \quad \dot{\psi}_2 = -\frac{\partial \hat{H}}{\partial x_2} = 2x_2 - \psi_1 + \psi_2 \, . \tag{9.41}$$

According to the transversality conditions we have $\psi_1(1) = \psi_2(1) = 0$. In the regions where $|u| < u_0$ it follows from the extremum condition $\partial \hat{H} / \partial u = 0$ that $\psi_2 = 2\alpha u$.

Solution for $u = 0$ shows that $\psi_2(t)$ is a decreasing function. Therefore, it is reasonable to assume that $u = u_0$ for $t \in [0, t_1]$ and $u < u_0$ for $t \in [t_1, 1]$. The value of t_1 is for the present unknown and will be calculated in the course of the solution.

Let us assign some value to t_1 and integrate the state equations for $t \in [0, t_1]$. According to the wavelet method we take (the matrices H and P_1 are calculated for $A = 0,\ B = t_1$):

$$\begin{aligned} \dot{x}_1 &= a_1 H\,, \quad & x_1 &= a_1 P_1\,, \\ \dot{x}_2 &= a_2 H\,, \quad & x_2 &= a_2 P_1 - E\,. \end{aligned} \tag{9.42}$$

Replacing these results into the state equations $\dot{x}_1 = x_2,\ \dot{x}_2 = -x_2 + u_0$ we find

$$\begin{aligned} & a_1 H - a_2 P_1 = -E\,, \\ & a_2(H + P_1) = (1 + u_0)E\,. \end{aligned} \tag{9.43}$$

Solving this system we find the wavelet coefficient vectors $a_1,\ a_2$ and calculate the functions $x_1,\ x_2$ according to (9.42). In the following we need the values $x_1^* = x_1(t_1),\ x_2^* = x_2(t_1)$. It follows from (2.6) that $p_1(t_1) = t_1$ and $p_i(t_1) = 0$ for $i \neq 1$. In view of (9.42) we find

$$x_1^* = a_1(1)t_1\,, \quad x_2^* = a_2(1)t_1 - 1\,. \tag{9.44}$$

Now let us consider the subinterval $t \in [t_1, 1]$. Again we divide this interval into $2M$ equal parts and calculate the matrices H, P and R from (2.1), (2.6), (2.7), (9.18) assuming $A = t_1,\ B = 1 - t_1$.

The solution is sought in the form

$$\begin{aligned} \dot{x}_1 &= \hat{a}_1 H\,, \quad & x_1 &= \hat{a}_1 P_1 + x_1^* E\,, \\ \dot{x}_2 &= \hat{a}_2 H\,, \quad & x_2 &= \hat{a}_2 P_1 + x_2^* E\,, \\ \dot{\psi}_1 &= \hat{b}_1 H\,, \quad & \psi_1 &= \hat{b}_1 R\,, \\ \dot{\psi}_2 &= \hat{b}_2 H\,, \quad & \psi_2 &= \hat{b}_2 R\,. \end{aligned} \tag{9.45}$$

Here $\hat{a}_1, \hat{a}_2, \hat{b}_1, \hat{b}_2$ denote wavelet coefficients for the subinterval $t \in [t_1, 1]$. The matrix R is calculated according to (9.18).

Substituting (9.45) into (9.39), (9.41) and taking into account that $\psi_2 = 2\alpha u$ we get the system

$$\begin{aligned} & \hat{a}_1 H - \hat{a}_2 P_1 = x_2^* E\,, \\ & \hat{a}_2(H + P_1) = \tfrac{1}{2\alpha}\, \hat{b}_2 R = -x_2^* E\,, \\ & -2\, \hat{a}_1 P_1 + \hat{b}_1 H = 2x_1^* E\,, \\ & -2\, \hat{a}_2 P_1 + \hat{b}_1 R + \hat{b}_2(H - R) = 2x_2^* E\,, \end{aligned} \tag{9.46}$$

which can be solved numerically with the aid of MatLab programs.

The control $u(t)$ must be continuous at $t = t_*$. In the case of an arbitrary chosen value t_*, such a requirement is not fulfilled. This discrepancy can be estimated by the function $\Delta = u(t_1 - 0) - u(t_1 + 0)$. Since

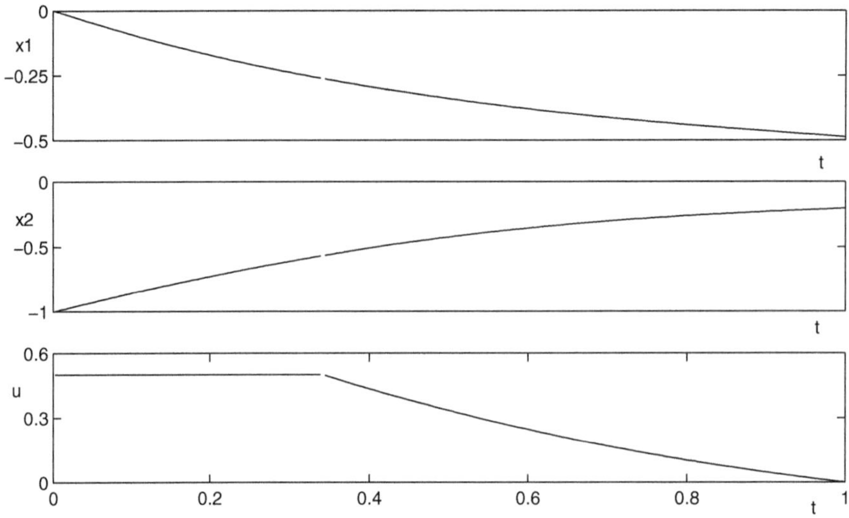

Fig. 9.1 Solution of the problem (9.39) for $u_0 = 0.5, \alpha = 0.5$

Table 9.2 Parameter t_1 and boundary values of the problem (9.39)

J	t_1	$x_1(1)$	$x_2(1)$
4	0.338569	−0.48679	−0.20527
5	0.338570	−0.48676	−0.20535
6	0.338574	−0.48675	−0.20536

$$u(t_1 - 0) = u_0 \,, \quad u(t_1 + 0) = \frac{1}{2\alpha}\psi_2(t_1 + 0) = \frac{1}{2\alpha}\hat{b}_2 R|_{t=t_1} = -\frac{1}{2\alpha}\hat{b}_2(1)(1 - t_1)$$

we obtain

$$\Delta = \frac{1}{2\alpha}\hat{b}_2(1)(1 - t_1) + u_0 \,. \tag{9.47}$$

We shall vary t_* until the condition $\Delta = 0$ is fulfilled with the necessary exactness.

Computer simulation was carried out for $u_0 = 0.5, \alpha = 0.5$; the results are plotted in Fig. 9.1. Since in the present case we do not know the exact solution, the error estimates (9.26) are unusable. Therefore we estimate the exactness of solution by calculating the values of t_1, $x_1(1)$, $x_2(1)$ at different levels of resolution J. These results are presented in Table 9.2.

* * *

Efficiency of the Haar wavelet method for solving optimal control problems is demonstrated. Numerical solutions for three test problems with different equality and inequality constrains are presented. Very satisfactory exactness of the results, even for a low number of collocation points is stated. In the case of some simple problems, the Haar wavelet method allows us to obtain analytical (exact) results (Sect. 9.4).

For simplicity of discussion, only linear problems were considered, but the recommended method of solution is applicable also for nonlinear systems. In this case, the wavelet coefficients must be calculated by some numerical technique, e.g. by the Newton method.

9.6 Related Papers

In several papers wavelet methods have been applied for solution of problems of the classical calculus of variations. Here the minimum of functional

$$J(x) = \int_0^1 F[t, x(t), \dot{x}(t)]dt \tag{9.48}$$

has to be found. This problem is reduced to solving the Euler-Lagrange equation

$$\frac{\partial F}{\partial x} - \frac{d}{dt}(\frac{\partial F}{\partial \dot{x}}) = 0 \tag{9.49}$$

with the appropriate boundary conditions. Different generalizations such as moving boundaries, integral constraints, etc. are possible.

For solving this variational problem, the function F is expanded into the wavelet series and the integral (9.48) is evaluated. The wavelet coefficients a_i are calculated from the extremality conditions $\partial J / \partial a_i = 0, \ i = 1, 2, \ldots$. This approach has been realized for different wavelets.

Chen and Hsiao [3] introduced for it the Walsh series. Walsh wavelets are also applied by Sloss and Blyth [23], by Glabisz [4]. Hsiao [6] presented a solution for the variational problem (9.48) via Haar wavelets, the method is illustrated by three numerical examples (one of them is the solution of the heat equation). The Haar approach was also used by Razzaghi and Ordokhani [19]. From other wavelet approaches we note here Legendre [2, 20, 21], Laguerre [11] and Chebyshev [5] methods.

Now let us discuss some papers in which optimal control theory problems are solved by the wavelet methods. Hsiao and Wang [7–9] investigated state analysis and optimal control of linear time-varying systems via Haar wavelets; the adjoint equations (9.4) were solved by the backward integration. The problem was solved via a single-term algorithm. Karimi [12] presented a computation method for optimal control of time-varying state-delayed systems

$$\begin{aligned} &\dot{x}(t) = A(t)x(t) + A_\alpha(t)(t-h) + B(t)u(t), \ t \in [0, 1], \\ &x(t) = \varphi(t), \ t \in [-h, 0], \ x(0) = x_0. \end{aligned} \tag{9.50}$$

A quadratic cost function

$$J = \frac{1}{2}x^T(1)Sx(1) + \frac{1}{2}\int_0^1 [x^T(t)Qx(t) + u^T(t)Ru(t)]dt \tag{9.51}$$

is to be minimized.

The problem was solved via Haar wavelets; into the Haar series are expanded the state function $x(t)$ and the control $u(t)$. Second-order linear systems were considered by Karimi et al. [13].

Linearly constrained quadratic optimal control problems were solved by the hybrid functions (consisting of block-pulse functions and Legendre polynomials) in [17, 22].

A numerical method for solving nonlinear control problems including state and control inequality constraints, based on the Haar wavelets, was proposed by Marzban and Razzaghi [17, 18]. In this paper the following optimal control problem is considered:

Minimize the functional

$$J = h(x(1), t_f) + \int_0^1 g[x(t), u(t), t, t_f]dt \tag{9.52}$$

subject to

$$\dot{x}(t) = f[x(t), u(t), t, t_f], t \in [0, 1], x(0) = x_0, s_i[x(t), u(t), t, t_f] \leq 0, i = 1, 2, \ldots, \nu. \tag{9.53}$$

In the course of the system, the inequality constraints are replaced by equality constraints

$$s_i[x(t), u(t), t, t_f] + z_i^2(t) = 0, \tag{9.54}$$

where $z_i(t)$ are auxiliary functions.

The functions $x(t),\ u(t),\ z_j(t)$ are expanded into the Haar series and the corresponding wavelet coefficients are calculated from the extremality conditions for the functional J.

We would also like to cite here also the paper by Tricaud and Chen [24] in which fractional order optimal control problems are solved, the solution is based on the MatLab toolbox program RIOTS95.

In all these papers, the operational matrix method, which we analyzed in Sect. 8.6, is applied. An essential difference between this method and our approach exists. In papers cited above, the control $u(t)$ was also expanded into the Haar series and the constraints $u_0 < u(t) < u_1$ were satisfied numerically by introducing auxiliary variables. In our approach, the controls $u(t)$ are calculated directly from conditions (9.5), (9.6). This allows us essentially to reduce the amount of numerical calculations; especially in the case where the control is bang-bang type (this means that in the whole time domain, the control can acquire only extremal values $u(t) = u_{max}$ or $u(t) = u_{min}$).

References

1. Bryson, A., Ho, Y.: Applied Optimal Control. Hemisphere, Washington (1975)
2. Chang, R., Wang, M.: Shifted Legendre direct method for variational problems. J. Optim. Theory Appl. **39**, 297–307 (1983)
3. Chen, C., Hsiao, C.: A Walsh series direct method solving variational problems. J. franklin Inst. **300**, 265–280 (1975)
4. Glabisz, W.: Direct Walsh-wavelet packet method for variational problems. Appl. Math. Comput. **159**, 769–781 (2004)
5. Horng, I., Chou, J.: Shifted Chebyshev direct method for variational problems. Int. J. Syst. Sci. **16**, 855–861 (1985)
6. Hsiao, C.: Haar wavelet direct method for solving variational problems. Math. Comput. Simul. **64**(5), 569–585 (2004)
7. Hsiao, C., Wang, W.: State analysis and optimal control of linear time-varying systems via Haar wavelets. Optimal Control Appl. Methods **19**, 423–433 (1998)
8. Hsiao, C., Wang, W.: Optimal control of linear time-varying systems via Haar wavelets. J. Optim. Theory Appl. **103**, 641–655 (1999)
9. Hsiao, C., Wang, W.: State analysis and optimal control of time-varying discrete systems via Haar wavelets. J. Optim. Theory Appl. **103**, 623–640 (1999)
10. Hull, D.: Optimal Control Theory for Applications. Springer, New York (2003)
11. Hwang, C., Shih, Y.: Laguerre series direct method for variational problems. J. Optim. Theory Appl. **39**, 143–149 (1983)
12. Karimi, H.: A computational method for optimal control problem of time-varying state-delayed systems by Haar wavelets. Int. J. Comput. Math. **83**, 235–246 (2006)
13. Karimi, H., Moshiri, B., Lohmann, B., Maralani, P.: Haar wavelet-based approach for optimal control of second-order linear systems in time domain. J. Dyn Control Syst. **11**, 237–252 (2005)
14. Lee, E., Marcus, L.: Foundations of Optimal Control Theory. Krieger, Malabar (1986)
15. Leitman, G.: The Calculus of Variations and Optimal Control. Plenum, New York (1981)
16. Lepik, Ü.: Solution of optimal control problems via Haar wavelets. Int. J. Pure Appl. Math. **55**, 81–94 (2009)
17. Marzban, H., Razzaghi, M.: Hybrid functions approach for linearly constrained quadratic optimal control problems. Appl. Math. Model. **27**, 471–485 (2003)
18. Marzban, H., Razzaghi, M.: Rationalized Haar approach for nonlinear constrained optimal control problems. Appl. Math. Model. **34**, 174–183 (2010)
19. Razzaghi, M., Ordokhani, Y.: An application of rationalized Haar functions for variational problems. Appl. Math. Comput. **122**(3), 353–364 (2001)
20. Razzaghi, M., Ordokhani, Y.: Solution of differential equations via rationalized Haar functions. Math. Comput. Simul. **56**(3), 235–246 (2001)
21. Razzaghi, M., Yousefi, S.: Legendre wavelets direct method for variational problems. Math. Comput. Simul. **53**(3), 185–192 (2000)
22. Razzaghi, M., Yousefi, S.: Legendre wavelets method for the solution of nonlinear problems in the calculus of variations. Math. Comput. Model. **34**, 45–54 (2001)
23. Sloss, B., Blyth, W.: A variational method using Walsh functions. Nonlinear Anal. **32**, 549–561 (1998)
24. Tricaud, C., Chen, Y.: An approximate method for numerically solving fractional order optimal control problems of general form. Comput. Math. Appl. **59**, 1644–1655 (2010)

Chapter 10
Buckling of Elastic Beams

10.1 Problem Statement and Method of Solution

In the previous chapters we have applied the wavelet approach for solving problems of mathematical calculus. In the subsequent chapters we will demonstrate that the Haar wavelet method is also a valuable tool in structural mechanics. To begin with, the buckling of elastic beams is discussed. The material in this chapter has been published by Lepik [18].

Consider an elastic Euler-Bernoulli beam with a rectangular cross-section. The beam's height is denoted by h, the width b and the length L. The beam is loaded by axial compressive force P. If the load attains the critical value P_{cr}, then the beam buckles. Our aim is to calculate buckling load P_{cr} and also the deflection curve of the buckled beam $w = w(\tilde{x})$.

The governing equation of the problem is

$$EI(\tilde{x})d^2w/d\tilde{x}^2 = M(\tilde{x}), \ \tilde{x} \in [0, L]. \tag{10.1}$$

Here E is the Young modulus, I is the moment of inertia of the cross-section; M denotes the bending moment at the cross-section $\tilde{x}$. In the following analysis two variants of the boundary conditions are considered:

$$\begin{aligned} &(i) \text{ beam with simply supported ends} : M = Pw, \\ &(ii) \text{ cantilever beam with clamped ends} : M = -P(\delta - w), \ \delta = w(L). \end{aligned} \tag{10.2}$$

Let us change the argument $x = \tilde{x}/L$. The moment of inertia for a rectangular cross-section is

$$I(x) = \frac{1}{12}bh^3(x). \tag{10.3}$$

By introducing the notations

DOI: 10.1007/978-3-319-04295-4_10,

$$\gamma(x) = \frac{h(x)}{h(0)}, \quad G(x) = \gamma^3(x), \quad \lambda = \frac{12L^2P}{EB[h(0)]^3} \tag{10.4}$$

Eq. (10.1) can be put into the form
(i) for the simply supported beam

$$Gw'' + \lambda w = 0, \quad w(0) = w(1) = 0, \tag{10.5}$$

(ii) for the cantilever beam

$$Gw'' + \lambda w = \lambda\delta, \quad w(0) = w'(0) = 0, \quad w(1) = \delta. \tag{10.6}$$

Here and in the following, primes denote differentiation with regard to x.

According to the Haar wavelet method the solution of (10.5) and (10.6) is sought in the form

$$w''(x) = \sum_{i=1}^{2M} a_i h_i(x); \tag{10.7}$$

Integrating this equation we obtain

$$w'(x) = \sum_{i=1}^{2M} a_i p_{1i}(x) + C_1,$$
$$w(x) = \sum_{i=1}^{2M} a_i p_{2i}(x) + C_1 x + C_2. \tag{10.8}$$

Here a_i are the wavelet coefficients; the quantities h_i, p_{1i}, p_{2i} are evaluated from (2.1), (2.6) and (2.7), respectively. The integration constants are calculated from the boundary conditions and we obtain

(i) $C_1 = -\sum_{i=1}^{2M} a_i q_2(i), \ C_2 = 0$ for the simply supported beam,
(ii) $C_1 = C_2 = 0$ for the cantilever beam.

All these results are substituted into (10.5) or (10.6) and the obtained differential equation is satisfied in the collocation points. By doing this we get the system of equations

$$\sum_{i=1}^{2M} a_i R(i,l) = \beta_2 F(l), \ l = 1, 2, \ldots, 2M, \tag{10.9}$$

where

$$R(i,l) = G(l)H(i,l) + \lambda[P_2(i,l) + \beta_1 Q(i,l)]. \tag{10.10}$$

Here the notations $Q(i,l) = -P_{2,i}(1)x_l, \ F(l) = \lambda\delta$ are introduced. For a simply supported beam we have $\beta_1 = 1, \ \beta_2 = 0$; for the cantilever beam $\beta_1 = 0, \ \beta_2 = 1$.

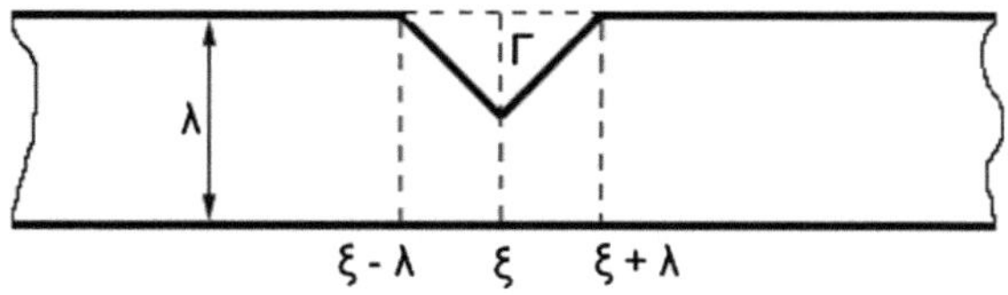

Fig. 10.1 *V*-shaped crack model

The system (10.9) can be put into the matrix form

$$aR = \beta_2 F. \tag{10.11}$$

This system is linear and homogeneous with regard to the variables a_i, δ. For getting a nontrivial solution the determinant of this system must be zero. From this requirement the critical load parameter λ_{cr} is evaluated.

To obtain the deflection curve $w = w(x)$, we must specify the deflection in some cross-section $x = x*$. According to (10.8) we get a complementary equation

$$w(x*) = \sum_{i=1}^{2M} a_i p_{2i}(x*) + C_1 x * + C_2, \tag{10.12}$$

which is incorporated into the system (10.11). The augmented system is now non-homogeneous and has a nontrivial solution for a_i and δ. The function $w = w(x)$ is calculated from (10.8).

10.2 Modelling Cracks

Cracks, occurring in structural elements, may significantly change the behavior of the whole structure and therefore the development of reliable crack models has been the objective of many investigations. Due to the great number of published papers, it is not possible to discuss all of them, therefore we confine ourselves only with a special crack model, which is described below. More complete analysis can be found in overviews, from which we note here Dimarogonas [9], Doebling et al. [10]. From more recent papers, we recommend Zhong and Oyadiji [26] and Caddemi et al. [3]; in these papers, a useful list of references is also added.

Our analysis is based on the *V*-shape crack model, which was proposed by Christides and Barr [6]. To simplify the solution in this Section, only beams of constant thickness $h = h_0$ are considered.

We assume that a crack occurs in some cross-section $x = \xi$. The crack has uniform depth d across the width of the beam and is fully open (Fig. 10.1). The relative crack depth is $\beta = d/h_0$. The quantity $G(\xi)$, characterizing the nondimensional rigidity of the cracked section, is

$$G(\xi) = (1 - \beta)^3. \tag{10.13}$$

Damage of cracked beam (i.e. decrease of rigidity) is

$$\Gamma = 1 - G(\xi) = 1 - (1 - \beta)^3. \tag{10.14}$$

According to Fig. 10.1 we have

$$G(x) = \begin{Bmatrix} 1 - (\Gamma/\Lambda)(x - \xi + \Lambda) & \text{for} & \xi - \Lambda \leq x \leq \xi, \\ 1 + (\Gamma/\Lambda)(x - \xi - \Lambda) & \text{for} & \xi \leq x \leq \Lambda + \xi, \\ 1 \;\; \textit{elsewhere}. & & \end{Bmatrix} \tag{10.15}$$

The parameter Λ denotes half of the crack width, the main problem is how to calculate it. Here two cases must be distinguished.

(i) The crack is considered as a notch or saw-cut. In this case Λ is prescribed from the designer (or from other considerations) and we can treat our problem as the buckling of a beam of variable thickness.

(ii) More complicated is the case of fatique cracks. Here different formulae for calculating Λ have been proposed.

Most of them can be written in the form

$$\Lambda = K(\beta)h_0/L. \tag{10.16}$$

According to Christides and Barr [6], a good approximation gives $K(\beta) = 1.5$. Bilello [1] recommended taking $K(\beta) = \beta/0.9$. Shen and Pierre [21] evaluated this coefficient by a finite element approach and suggested the value $K(\beta) = 1.936^{-1} = 0.512$. A more complicated formula belongs to Bovsunovski and Matveev [2], which has the form

$$K(\beta) = \frac{0.3675(1 - \beta)}{1 - (1 - \beta)^3}[(1 - \beta)^6 - 3(1 - \beta)^2 + 2] \tag{10.17}$$

The calculation of Λ can be avoided if we approximate (14.11) by an expression involving an exponential function [22]:

$$G(x) = \frac{1 - \Gamma}{1 - \Gamma + \Gamma \exp[-2\alpha|x - \xi|L/h]}. \tag{10.18}$$

The value $\alpha = 0.667$ is estimated from experimental data.

So we can state that there are great discrepancies between the values of $K(\beta)$, predicted by different approaches.

In the following sections some problems about buckling of elastic beams are solved.

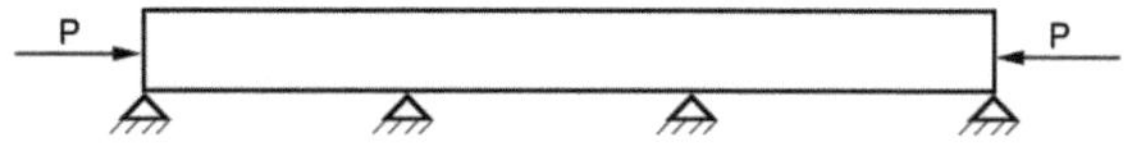

Fig. 10.2 Simply supported beam on two rigid intermediate supports

10.3 Beam on Intermediate Supports

Consider an axially compressed beam with n rigid supports at the cross-sections $y_\alpha \in (0, 1), \alpha = 1, 2, \ldots, n$. The conditions $w(y_\alpha) = 0$ can be interpreted as intermediate boundary conditions for the system (10.11) and in view of (10.12) we have

$$\sum_{i=1}^{2M} a_i p_{2i}(y_\alpha) + C_1 y_\alpha + C_2 = 0, \tag{10.19}$$

where the quantities $p_{2i}(y_\alpha)$ are calculated from (2.6) to (2.7). Adding these equations to the system (10.11) we get an augmented system of $2M + n$ equations. For getting non-trivial solutions, the rank of this system must be less than $2M$. For realizing this requirement, we choose from the augmented matrix a submatrix of order $2M$ and vary the load parameter so that the determinant of this submatrix turns to zero. Since the system is homogeneous, then all $2M$-order determinants made up from the augmented system are also zero. Consequently the calculated value for λ is really critical.

Example 10.1: Consider a simply supported beam on two rigid intermediate supports with the locations $y_1 = 1/3,\ y_2 = 2/3$ (Fig. 10.2). The exact solution of this problem is $\lambda_{cr} = 9\pi^2 = 88.82$ ([24], Sect. 19). Our computations gave $\lambda_{cr} = 89.14$ for $J = 4$ (with the error 0.3 %) and $\lambda_{cr} = 88.91$ for $J = 5$ (error 0.09 %). For putting together the deflection curve it is assumed $w'(0) = 1$. In view of (10.8) this condition gets the form

$$\sum_{i=1}^{2M} a_i q_2(i) = -1 \tag{10.20}$$

In view of (2.6) and (2.7) we find

$$q_2(i) = p_{2,i}(1) = \begin{cases} 0.5 \text{ for } i = 1, \\ 0.25(1/m)^2 \text{ for } i \neq 1. \end{cases} \tag{10.21}$$

According to the conventional method of solution, the multi-span beams are divided into spans between the two consequent supports. The governing equation is integrated separately for each span. The integration constants are calculated from the continuity conditions at the supports (consult e.g. [19, 24]). If the number of intermediate supports is greater than 2, this approach may turn out to be very troublesome. Our method is much more simple since we treat the beam as a whole (not dividing it into parts).

Table 10.1 Reduction of the buckling load of a simply supported beam with a crack at $x = 0.5$

β	0.2	0.4	0.6	0.8	0.9	0.95	0.99
λ_{cr}	9.72	9.47	9.01	7.83	6.14	4.23	1.18
Δ	0.99	0.96	0.91	0.79	0.62	0.43	0.11

10.4 Buckling of Cracked Beams

Assume that the moment of inertia of the cross-section is $I = I_0 = const$, except some discrete points $x = y_\alpha,\ \alpha = 1, 2, \ldots, n$, in which $I = I_\alpha < I_0$. Such a situation can be interpreted as a damaged beam carrying n cracks. If we apply the V-shaped crack model, then our problem is reduced to calculating the critical load of the beam with variable thickness. The nondimensional rigidity $G(x)$ is $G(x) = 1$ except in the vicinity of the crack sections where it is calculated according to (10.15). Next the systems of Eqs. (10.9) or (10.11) are solved by the method presented in Sect. 10.2.

Let us consider some examples; in these, the nondimensional width of the cracks takes the value $2\Lambda = 0.032$.

Example 10.2: Computer simulation was carried out for a simply supported beam with a crack at $x = 0.5$. The results for $J = 4$ are presented in Table 10.1. The symbol $\Delta = \lambda_{crack}/\lambda$ shows the relative reduction of the buckling capacity for the cracked beam. The functions $w(x),\ w'(x)$ and $w''(x)$ for some values of β are plotted in Fig. 10.3.

Example 10.3: Karaagac et al. [16] and Skrinar [23] investigated buckling of a cantilever beam with a rectangular cross-section. The beam has a crack located at $y \in (0, 1)$. The crack's depth is $\beta = 1 - h/h_0$, where h_0 is the beam's thickness, h—thickness of the damaged section. The crack is modelled by a linear massless rotational spring connecting the uncracked parts of the beam. The governing equation is integrated separately for both uncracked segments. For the connection between both parts, the continuity of displacement, bending moment and shear force is imposed.

It is interesting to compare these results with our outcome. We have carried out computations for three variants of the crack location $y_1 = 0.2,\ y_2 = 0.5,\ y_3 = 0.8$. For the crack depth the value $\beta = 0.5$ was taken. The two crack parameters are related according to the formula $\Gamma = (1 - \beta)^3$; in the present case $\Gamma = 0.875$. An overview of the obtained results is presented in Table 10.2. The symbol Δ_w denotes our wavelet results. Experimental data from the paper [16] are marked by Δ_{exp}. The coefficients Δ_1 correspond to a FEM solution from the same paper. Data in the last two columns are taken from the paper by [23]. For calculating Δ_2, the COSMOS 2D FEM program was applied (here 20,000 8-noded finite elements with more than 122,000 degrees of freedom were used). The coefficients Δ_3 are calculated by the analytical model of Skrinar.

It follows from Table 10.2 that results obtained by the Haar wavelet method are in satisfactory accordance with the data of other authors. Our method is substantially simpler, since we do not need to introduce rotational springs and divide the beam into

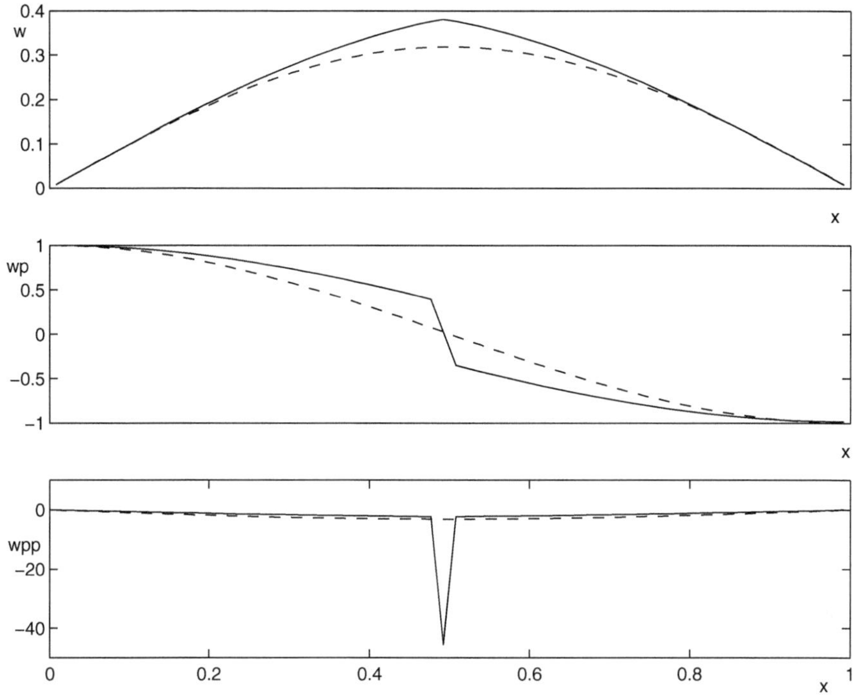

Fig. 10.3 Deflection w, slope wp and curvature wpp of a simply supported beam with a crack at $x = 0.5$; *Solid lines* damaged beam, *doted lines* undamaged beam

Table 10.2 Reduction of the buckling load of a cracked cantilever beam $\Delta = \lambda_{crack}/\lambda$, comparison of theoretical and experimental results

y	Δ_w	Δ_{exp}	Δ_1	Δ_2	Δ_3
0.2	0.832	0.789	0.904	0.900	0.891
0.5	0.900	0.915	0.978	0.933	0.937
0.8	0.980	0.978	0.989	0.985	0.987

parts between the damaged sections. Difference is also in continuity conditions at the cracked sections. In the conventional approach, the gradient w' is regarded as discontinuous. In our solution, it is continuous (discontinuous is the curvature). This variant of the continuity conditions seems to us more logical, since if the first derivative w' is discontinuous then the second derivative is indeterminate. We consider our results about crack simulation as preliminary—this model should be tested by solving more problems about cracked structures. This is the object of further research.

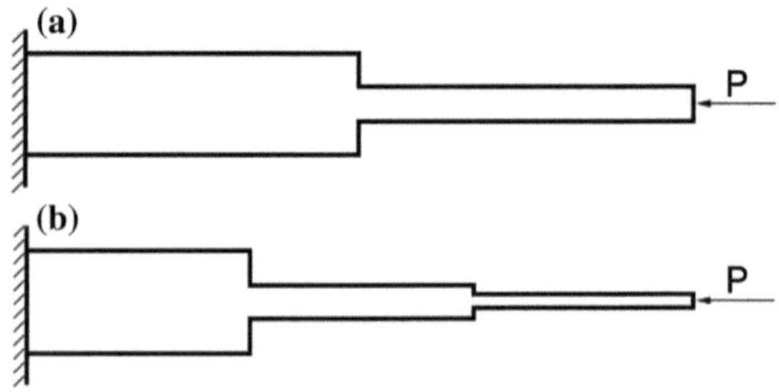

Fig. 10.4 Buckling of two- and three stepped cantilever beams

10.5 Buckling of Beams of Variable Cross-Section

Consider the case where the moment of inertia $I(x)$ is a given piecewise constant function of the coordinate x. Let us assume $I(0) \neq 0$ and denote $G(x) = I(x)/I(0)$. The function $G(x)$ may have points of discontinuity $x = y_\alpha,\ \alpha = 1, 2, \ldots, n$. We have to find such a value of λ at which the buckling begins.

In the text-book [24], Sect. 26, buckling of cantilever beams of variable cross-section is discussed. With the purpose to compare our results with the Timoshenko's data we take

$$G(x) = G_1[1+(G_1^{-1/n} - 1)(1-x)]^n,\ G_1 = I(1)/I(0),\ n = 1, 2, 3, \ldots \quad (10.22)$$

If $n = 2$ the exact solution of the problem exists. For $n > 2$ the solution can be found with the aid of the Bessel functions. Timoshenko's numerical results for $n = 2$ and $n = 4$ are presented on Tables 11 and 12 of his book. We have carried out computations for $J = 5$, $G_1 = 0.4$. In the case $n = 2$ we have $\lambda_{cr} = 1.906$ (Timoshenko's solution gives 1.904). For $n = 4$ both solutions gave the same result $\lambda_{cr} = 1.870$.

Example 10.4: The proposed method of solution is applicable also for stepped beams (in this case $G = G(x)$ is a piecewise constant function). Consider a two-stepped beam with $y_1 = 0.4$ and $G(x) = 1$ for $x \in [0, y_1],\ G(x) = 0.6$ for $x \in [y_1, 1]$ (Fig. 10.4a). Computer simulation for $J = 4$ gave $\lambda_{cr} = 2.056$. The same problem was solved by Timoshenko, Sect. 25 [24], who obtained $\lambda_{cr} = 2.048$. The difference between the both solutions is 0.39 %.

Computations were carried out also for the three-stepped beam for which $y_1 = 1/3,\ y_2 = 2/3$ and $G(x) = 1$ for $x \in [0, y_1],\ G(x) = 0.7$ for $x \in [y_1, y_2],\ G(x) = 0.4$ for $x \in [y_2, 1]$ (Fig. 10.4b). The critical load parameter value $\lambda_{cr} = 1.988$ was obtained. Again we can state that our solution is essentially simpler when compared with the traditional approaches, since we treat the beam as a whole and do not divide it into parts for which $I(x) = \text{const}$.

10.6 Buckling of Beams on Elastic Foundation

The governing equation for buckling and vibrations of an axially compressed beam on elastic foundation is

$$EI\frac{\partial^4 w}{\partial \tilde{x}^4} + (P - \tilde{K}_2)\frac{\partial^2 w}{\partial \tilde{x}^2} + \tilde{K}_1(\tilde{x})w = -\rho A\frac{\partial^2 w}{\partial t^2}. \qquad (10.23)$$

Here it is assumed that I = const, $\tilde{K}_1(\tilde{x})$ is the variable coefficient of Winkler foundation, $\tilde{K}_2$ the Pasternak foundation coefficient, ρ is the material density of the beam, A—the cross-sectional area.

Introducing dimensionless quantities

$$x = \tilde{x}/L,\ \lambda = PL^2/(EI),\ K_1 = \tilde{K}_1 L^4/(EI),\ \tilde{K}_2 = \tilde{K}_2 L^2/(EI),\ \mu = \rho A L^4/(EI) \qquad (10.24)$$

we can rewrite (10.23) in the dimensionless form

$$w^{IV} + (\lambda - K_2)w'' + K_1(x)w = -\mu w^{\cdot\cdot}. \qquad (10.25)$$

Solution of this equation is sought in the form

$$w(x,t) = W(x)T(t). \qquad (10.26)$$

Replacing (10.26) into (10.25) and separating the variables we obtain

$$T(t) = Acos(\omega t) + Bsin(\omega t), \qquad (10.27)$$

$$W^{IV} + (\lambda - K_2)W'' + [K_1(x) - \mu\omega^2]W = 0, \qquad (10.28)$$

where ω denotes the frequency of the beam vibrations.

The wavelet solution of (10.28) is sought in the form

$$W^{IV} = aH. \qquad (10.29)$$

By multiple integration of this equation we find

$$\begin{aligned}
W''' &= aP_1 + C_1 E,\\
W'' &= aP_2 = C_1 x + C_2 E,\\
W' &= aP_3 + (1/2)C_1 x.^2 + C_2 x + C_3 E,\\
W &= aP_4 + (1/6)C_1 x.^3 + (1/2)C_2 x.^2 + C_3 x + C_4 E.
\end{aligned} \qquad (10.30)$$

Here $E = [1, 1, \ldots, 1]$ and the decimal point denotes element-by element multiplication. The integration constants $C_1 - C_4$ are calculated from the boundary conditions:

(i) For a simply supported beam the boundary conditions are $W(0) = W''(0) = W(1) = W''(1) = 0$ and we obtain

$$C_1 = -aq_2, \; C_2 = C_4 = 0, \; C_3 = -a[q_4 - 0.5q_2]. \tag{10.31}$$

(ii) Satisfying the boundary conditions for the cantilever beam $W(0) = W'(0) = W''(1) = W'''(1) = 0$ we get

$$C_1 = -aq_1, \; C_2 = a(q_2 - q_1), \; C_3 = C_4 = 0. \tag{10.32}$$

The vectors $[q_\alpha(i)] = [p_{\alpha i}(1)], \; \alpha = 1, 2, 3, 4$ are calculated according to (2.6) and (2.7). All these results are replaced into (10.28) and the obtained equation is satisfied in the collocation points. The outcome can be presented in the matrix form as $aR = 0$, where

$$\begin{aligned} R(i,l) &= H(i,l) + (\lambda - K_2)[P_2(i,l) + x_l q_2(i)] \\ &\quad + [K_1(l) - \mu\omega^2]\{P_4(i,l) + x(l)(q_2(i)/6 - q_4(i) - x_l^3 q_2(i)/6]\} \end{aligned} \tag{10.33}$$

for the simply supported beam and

$$\begin{aligned} R(i,l) &= H(i,l) + (\lambda - K_2)[P_2(i,l) - x_l q_2(i) + q_2(i) - q_1(i)] \\ &\quad + [K_1(l) - \mu\omega^2]\{P_4(i,l) + x(l)^2[-x_l q_1(i) + 3q_2(i) - q_1(i)]/6\} \end{aligned} \tag{10.34}$$

for the cantilever beam.

Such critical values for λ (or ω), for which the determinant $|R|$ is zero, must be found.

Example 10.5: Calculate the buckling load for an axially compressed simply supported beam on Winkler's foundation. This problem was discussed in [24] in Sect. 21. Timoshenko gave for the critical load the formula, which in our notations has the form

$$\lambda_{cr} = \pi^2(m + \frac{K_1}{m^2\pi^4}), \quad K_1 = const, \tag{10.35}$$

Here m is a positive integer. For $K_1 = 100$ it follows from (10.35) that $\lambda_{cr} = 20.003$ and $m = 1$; our wavelet solution gives $\lambda_{cr} = 20.007$. For $K_1 = 400$ Timoshenko's solution is $\lambda_{cr} = 49.61, \; m = 2$; our solution gives $\lambda_{cr} = 49.62$. So we see that accordance of the two solutions is rather good. The displacement curves for both solutions are plotted in Fig. 10.5.

Example 10.6: Chen [4] considered free vibrations of a cantilever beam on Winkler's foundation. First three critical frequencies $\omega_1, \; \omega_2, \; \omega_3$ were calculated by the differential quadrature element method (DQEM). Computer simulation was carried out for $\lambda = 0, \; K_2 = 0, \; K_1 = 1, \; \mu = 1$ (our notations).

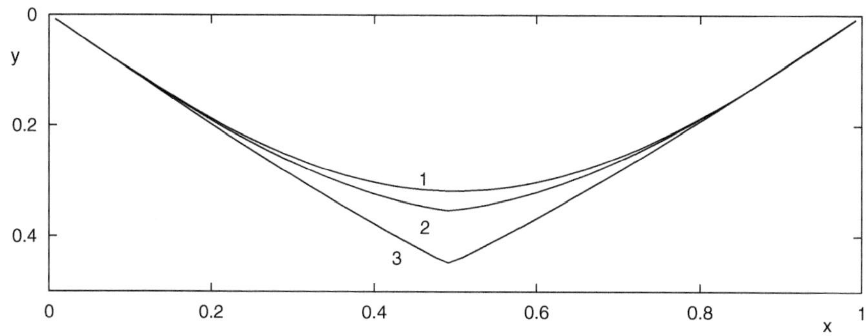

Fig. 10.5 Buckling of a simply supported beam on Winkler's foundation. *1* solution for $K_1 = 0$, *2* solution for $K_1 = 100$, *3* solution for $K_1 = 400$

Table 10.3 Critical frequencies of a vibrating cantilever beam on Winkler's foundation

ω	Exact	Wavelet	DQEM1	DQEM2
ω_1	3.655	3.656	3.655	3.655
ω_2	22.056	22.063	22.056	22.057
ω_3	61.706	61.731	61.668	61.706

We have solved the same problem by the Haar wavelet method for $J = 4$ (32 collocation points). All these results are shown in Table 10.3. Since in the present case (10.28) is a linear equation with constant coefficients, it is not difficult to put together an exact solution of the problem. For conciseness sake, the course of this solution is not shown here; the calculated critical frequencies are indicated in the second column of Table 10.3. In the third column our wavelet results are presented. The data in the fourth and fifth column are taken from Chen's paper [4]: the data DQEM1 correspond to five-node model with eight elements; for DQEM2 the nine-mode model with eight elements elements was used. A good accordance of all these results can be stated.

10.7 Related Papers

Solving problems of structural mechanics by the wavelet methods has been the subject of a number of investigations within the last two decades. Different wavelet families have been applied. Zhou and Zhou [28] made use of the Daubechies and Coiflet wavelets. Chaotic states of viscoelastic beams were investigated with the aid of the Walsh wavelet packet by Glabisz [14]. The wavelet approach is often combined with the finite element method [5, 8, 15].

Plane elasticity problems were investigated with the aid of two-dimensional interpolating wavelets by Kim et al. [17]. B-spline wavelets on the interval were applied for exploring stability of elastic plates and shells and also the 3D-problems were applied by Zhong and Xiang [27].

In several papers the wavelet methods have been applied for crack detection and damage estimation in elastic beams. An overview about such contributions can be found in a paper by Gentile and Messina [13]. In addition to it we recommend the following papers. Zhong and Oyadiji [26] used for crack detection stationary wavelet transform. Deng and Wang [7] applied the discrete wavelet transform to locate a crack along the length of the beam. Wang and Deng [25] extended the analysis to a plate with a through-thickness crack. In the last paper, the Haar wavelets were used with success. In these papers the authors were able to detect relatively small cracks, but a method for estimating the crack extent has not been proposed. Both, the location and size of the cracks were estimated by Douka et al. [11] who applied for this purpose the "symmetrical 4" wavelet. A 2D continuous wavelet transform for damage detection of plate structures was used by Fan and Qiao [12].

Considering the subject matter of the present book our main interest is focused on papers in which the solutions are based upon the Haar wavelets. Quek et al. [20] applied the Haar and Gabor wavelets for crack detection in beams; the authors state that for detection of cracks, Haar wavelets exhibit superior performance. Kim et al. [17] introduced a vibration-based damage evaluation technique based on the Haar wavelets.

References

1. Bilello, E.: Theoretical and experimental investigation of damaged beam under moving system. Dissertation, Universita degli Studi di Palermo (2001)
2. Bovsunovski, A., Matveev, V.: Analytical approach to the determination of dynamic characteristics of a beam with a closing crack. J. Sound Vib. **235**, 415–434 (2000)
3. Caddemi, S., Calio, I., Marietta, M.: The non-linear dynamic response of the Euler-Bernoulli beams with an arbitrary number of switching cracks. Int. J. Non-Linear Mech. **45**, 714–726 (2010)
4. Chen, C.: Vibrations of prismatic beams on an elastic foundation by the differential quadrature element method. J. Comput. Struct. **77**, 1–9 (2000)
5. Chen, X., Xiang, J., Li, B., He, Z.: A study of multiscale wavelet-based elements for adaptive finite elements analysis. Adv. Eng. Softw. **41**, 196–205 (2010)
6. Christides, S., Barr, A.: One dimensional theory of cracked Bernoulli-Euler beams. Int. J. Non-Linear Mech. **26**, 639–648 (1984)
7. Deng, X., Wang, Q.: Crack detection using spatial measurements and wavelet. Int. J. Fract. **91**, L23–L28 (2008)
8. Díaz, L., Martín, M., Vampa, V.: Daubechies wavelet beam and plate finite elements. Finite Elem. Anal. Des. **45**, 200–209 (2009)
9. Dimarogonas, A.: Vibration of cracked structures: a state of the art review. Eng. Fract. Mech. **55**, 831–857 (1996)
10. Doebling, S., Farrar, C., Prime, M.: A summary review of vibration-based damage identification methods. Shock Vib. Dig. **30**, 91–105 (1998)
11. Douka, E., Loutridis, S., Trochidis, A.: Crack identification in beams using wavelet analysis. Int. J. Solids Struct. **40**, 3557–3569 (2003)

12. Fan, W., Qiao, P.: A 2-D continuous wavelet transform of mode shape data for damage detection of plate structures. Int. J. Solids Struct. **46**, 4379–4395 (2009)
13. Gentile, A., Messina, A.: On the continuous wavelet transforms applied to discrete vibrational data for detecting open cracks in damaged beams. Int. J. Solids Struct. **40**, 295–315 (2003)
14. Glabisz, W.: The use of Walsh-wavelet packets in linear boundary value problems. Comput. Struct. **82**, 131–141 (2004)
15. He, W., Ren, W.: Finite element analysis of beam structures based on trigonometric wavelet. Finite Elem. Anal. Des. **51**, 59–66 (2012)
16. Karaagac, C., Örtürk, H., Sabuncu, M.: Free vibrations and lateral buckling of a cantilever slender beam with an edge crack: experimental and numerical studies. J. Sound Vib. **326**, 235–250 (2009)
17. Kim, B., Kim, H., Park, T.: Nondestructive damage evaluation of plates using the multi-resolution analysis of two-dimensional Haar wavelet. J. Sound Vib. **292**, 82–104 (2006)
18. Lepik, Ü.: Buckling of elastic beams by the Haar wavelet method. Est. J. Eng. **17**, 271–284 (2011)
19. Lin, H., Chang, S.: Free vibration analysis of multi-span beams with intermediate flexible constraints. J. Sound Vib. **281**, 155–169 (2005)
20. Quek, S., Wang, Q., Zhang, L., Ang, K.: Sensitivity analysis of crack detection in beams by wavelet technique. Int. J. Mech. Sci. **43**, 2899–2910 (2001)
21. Shen, M., Pierre, C.: Natural modes of Bernoulli-Euler beams with symmetric cracks. J. Sound Vib. **138**, 115–134 (1990)
22. Sinha, J., Friswell, M., Edwards, S.: Simplified models for the location of cracks in beam structures using measured vibration data. J. Sound Vib. **251**, 13–38 (2002)
23. Skrinar, M.: On the application of a simply computational model for slender transversely cracked beams in buckling problems. Comput. Mater. Sci. **39**, 242–249 (2009)
24. Timoshenko, S.: Theory of Elastic Stability. McGraw-Hill Book Company, New York (1936)
25. Wang, Q., Deng, X.: Damage detection with spacial wavelets. Int. J. Solids Struct. **36**, 3443–3468 (1999)
26. Zhong, S., Oyadiji, S.O.: Crack detection in simply supported beams without baseline modal parameters by stationary wavelet transform. Mech. Syst. Sign. Process. **21**, 1853–1884 (2007)
27. Zhong, Y., Xiang, J.: Construction of wavelet-based elements for static and stability analysis of elastic problems. Acta Mech. Solida Sin. **24**, 355–364 (2011)
28. Zhou, Y., Zhou, J.: A modified wavelet approximation of deflections for solving PDEs of beams and square thin plates. Finite Elem. Anal. Des. **44**, 773–783 (2008)

Chapter 11
Vibrations of Cracked Euler-Bernoulli Beams

11.1 Governing Equations

In this chapter, the Haar wavelet method is applied for analysing bending and vibrations of elastic Euler-Bernoulli beams. It is assumed that the beam has a rectangular cross-section and carries a concentrated load at $\tilde{x} = \tilde{x}_*$.

The equations of motion are

$$\frac{\partial \tilde{M}}{\partial \tilde{x}} = -\tilde{Q}, \quad \tilde{M} = -EI(\tilde{x})\frac{\partial^2 \tilde{w}}{\partial \tilde{x}^2}, \quad I(\tilde{x}) = (1/12)bh(\tilde{x})^3,$$
$$\frac{\partial \tilde{Q}}{\partial \tilde{x}} = -\mu\frac{\partial^2 \tilde{w}}{\partial t^2} + \tilde{F}\delta(\tilde{x} - \tilde{x}_*), \quad \mu = bh(\tilde{x})\rho. \tag{11.1}$$

Here E is the Young modulus, $I(\tilde{x})$—the moment of inertia of the cross-section, $\tilde{M}$ and $\tilde{Q}$ denote the bending moment and shear force, respectively. The beam's length is L, dimensions of its cross-section are $b,\ h(\tilde{x})$; ρ denotes the density of the beam material, $\delta(\tilde{x} - \tilde{x}_*)$ is the Dirac's delta function.

In the following, only motion with an harmonic load $\tilde{F} = F\exp(i\omega t)$ is considered. By introducing the symbols

$$w = \tilde{w}\exp(-i\omega t),\ M = \tilde{M}\exp(-i\omega t),\ Q = \tilde{Q}\exp(-i\omega t),\ x = \tilde{x}/L \tag{11.2}$$

we find (primes denote differentiation with regard to x)

$$M' = -LQ, \quad M = -EI(x)w''/L^2, \quad Q' = \mu L\omega^2 w + F\delta(x - x_*). \tag{11.3}$$

In view of (11.3) the governing equation can be put into the form

$$(G(x)w'')'' = -A\gamma\omega^2 w + BF\delta(x - x_*), \tag{11.4}$$

Ü. Lepik and H. Hein, *Haar Wavelets*, Mathematical Engineering,

DOI: 10.1007/978-3-319-04295-4_11,

where

$$\gamma(x) = h(x)/h(0), \quad G(x) = \gamma(x)^3, \quad h_0 = h(0) \tag{11.5}$$

and

$$A = \frac{12\rho L^4}{E h_0{}^2}, \quad B = \frac{12}{Eb}\left(\frac{L}{h_0}\right)^3. \tag{11.6}$$

Integrating (11.4) twice we find

$$Gw'' - \Omega \int_0^x dx \int_0^x \gamma w dx = BF(x - x_*)U(x - x_*) + C_1 x + C_2. \tag{11.7}$$

Here $\Omega = A\omega^2$ and U denotes the Heaviside function $U = 0$ for $x < x_*$ and $U = 1$ for $x \geq x_*$. The integration constants C_1, C_2 are calculated from the boundary conditions for M and Q. In the following, only the case of a cantilever beam, for which $M(1) = Q(1) = 0$, is considered. Besides, it is assumed that the load F acts in the end section $x = 1$. In this case, the integro-differential equation (11.7) obtains the form

$$Gw'' + \Omega\left[-\int_x^1 dx \int_0^x \gamma w dx + (1 - x)\int_0^1 \gamma w dx\right] = BF(1 - x). \tag{11.8}$$

If cracks exist then they are modelled as open V-shaped edge cracks (consult Sect. 10.2).

Equation (11.8) is solved by the Haar wavelet method. We introduce the vector of the wavelet coefficients $a = (a_i),\ i = 1, 2, \ldots, 2M$ and define the Haar matrices

$$H(i,l) = h_i[x_c(l)],\ P_\alpha(i,l) = p_{\alpha,i}[x_c(l)],\ i, l = 1, 2, \ldots, 2M, \alpha = 1, 2, 3, \ldots \tag{11.9}$$

where the symbols $x_c(l)$ denote the coordinates of the collocation points. The solution is sought in the matrix form $w'' = aH$. By integrating this equation and taking into account the boundary conditions we find

$$w' = aP_1, \quad w = aP_2. \tag{11.10}$$

Next we define the notations

$$I_{1i}(x) = \int_0^x \gamma(x) p_{2i}(x) dx, \quad I_{2i}(x) = \int_0^x dx \int_0^x \gamma(x) p_{2i}(x) dx, \tag{11.11}$$

These results are replaced into (11.8); the outcome can be put into the matrix form

$$aR = BF(E - x_c), \tag{11.12}$$

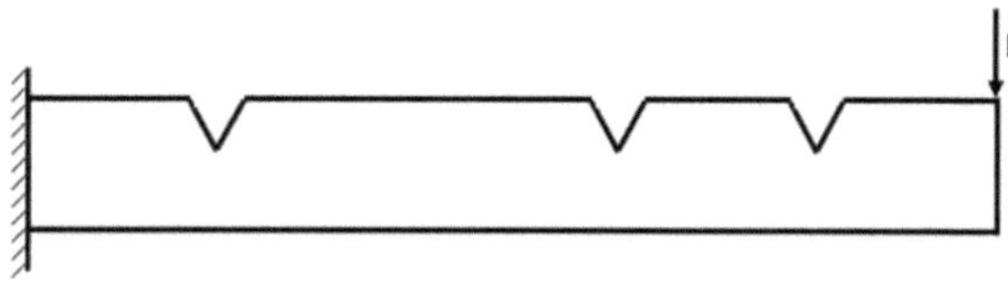

Fig. 11.1 Bending of a three-cracked cantilever beam

where

$$R(i,l) = G(l)H(i,l) + \Omega[E(l) - x_c(l)]I_{1i}(1) - I_{2i}(1) + I_{2i}[x_c(l)] \quad (11.13)$$

Here E is a $2M$ unit vector. The wavelet coefficients a_i are calculated by solving the system (11.12), the slope and deflection are computed from (11.10).

To demonstrate the efficiency of the proposed method in the following sections three problems are solved.

11.2 Bending of Multi-cracked Beams

Example 11.1: In the paper [30] by Skrinar, the following problem was solved by the finite element method. A cantilever beam of length $L = 8\,\text{m}$ and height $h = 0.18\,\text{m}$ carries a concentrated load $F = 10\,\text{kN}$ at the beam's end. The Young's modulus is $E = 30\,\text{GPa}$. At the distances 1, 4 and 6 m, three cracks are applied. The depths of all cracks were taken to be 0.09 m (Fig. 11.1). Let us solve this problem by the Haar wavelet method. We start from (11.12)–(11.13) by assuming $\Omega = 0$. In the present case $B = 0.3512$. The dimensionless crack locations and depths are $\xi_1 = 0.125$, $\xi_2 = 0.50$, $\xi_3 = 0.75$, $\beta_1 = \beta_2 = \beta_3 = 0.5$. The function $G(x)$ is calculated according to (10.15). For the level of resolution the value $J = 7$ was taken. Computer simulation was carried out for Bilello model $\Lambda_B = (\beta/0.9)(h/L)$ and for Christides-Barr model $\Lambda_C = 1.5h/L$. Deflection at the beam end was found to be 12.71 and 15.40, respectively (Skrinar got the value 14.42).

Discrepancy between the coefficients Λ_B and Λ_C exists also in the case of other problems; e.g. for calculating eigenfrequencies of vibrations. Bilello criterion gives mostly overestimated and Christides-Barr underestimated values.

To improve the solutions we propose a new criterion. We demand that the damaged areas for the functions $\gamma = \gamma(x)$ and $G = G(x)$ are equal. If we denote the half of the effective crack width for $\gamma = \gamma(x)$ by Λ_*, then the equality of the triangular damaged areas gives $\Lambda_*\beta = \Lambda_B\Gamma$ and

$$\Lambda_* = \frac{\Gamma}{\beta}\Lambda_B = \frac{\Gamma}{0.9}\frac{h_0}{L}. \quad (11.14)$$

Since $\beta \le \Gamma \le 1$ we have $\Lambda_B \le \Lambda_* \le \Lambda_C$, so our criterion gives intermediate values compared with the other two criteria.

Table 11.1 Deflections of the beam end (in cm)

	First crack	Second crack	Third crack	Under the force
Skrinar	0.263	4.483	9.097	14.417
Wavelet	0.219	4.430	8.980	14.223

It should be mentioned that the criterion (11.14) does not have any strict theoretical validity, it can be treated as an empirical coefficient, whose validity must be proved by comparing the results with experimental data and with the results of other theoretical papers. It follows from the calculations carried out in Sects. 11.2–11.4 that the results obtained with the aid of the criterion (11.4) are in satisfactory accordance with the data of other researchers. It should be once more emphasized that the problems for calculating the effective crack width appear only in the case of fatigue cracks; for saw-cut cracks, the value Λ is prescribed.

Now let us return to our problem. Calculations were repeated for Λ_* and the results are presented in Table 11.1, where comparisons with the Skrinar's results are given.

It follows from this table that the accordance of both solutions may be regarded as satisfactory. The Skrinar's computation model consisted of 14,400 2D eight noded finite elements and about 90,000 linear equations were solved. Our method is much simpler, since for $J = 7$ only 256 equations had to be solved.

11.3 Free Vibrations of Beams with Singularities

In the case of free vibrations we take $F = 0$ in (11.12). This system is now linear and has nontrivial solutions only if the determinant $\mid D \mid = 0$. This requirement is fulfilled for some discrete values $\Omega = \Omega_\nu,\ \nu = 1, 2, 3...$ The eigenfrequencies of the beam are $\omega_\nu = \sqrt{\Omega_\nu / A}$, where the coefficient A is calculated from (11.6). If we want the deflection modes, then we must fix the deflection at some point, assuming e.g. $w(1) = 1$ and adding this requirement to the system (11.12). The augmented system is non-homogeneous and can be solved in the ordinary way. If cracks appear, the non-dimensional rigidity in the damaged zone is calculated according to (10.15).

For simplicity sake, let us confine to a beam of constant thickness $\gamma(x) = 1$ which carries a crack at $x = \xi$. In the cracked zone $\xi - \Lambda < x < \xi + \Lambda$ we have $\gamma(x) = \sqrt[3]{G(x)}$. Next we have to evaluate the integrals (11.11), making use of (11.13). This process is quite troublesome therefore we do some simplifications. Since the damaged area in the beam is quite small the numerical values of these integrals do not change much if we presume that the equality $\gamma(x) = 1$ holds also in the cracked zone. In this case, we obtain simple formulae for calculating these integrals: $I_{1i}(x) = p_{3i}(x),\ I_{2i}(x) = p_{4i}(x)$. Essential is the choice of the parameter Λ, which determines the extent of the cracked zone. We have made use of the estimate $\Lambda = \Gamma h/(0.9L)$, which was discussed in Sect. 11.2.

Table 11.2 Reduction of eigenfrequencies of cracked beams, comparison of the results with the results of Sinha et al. [29]

β	Mode 1			Mode 2		
	Δ_{theor}	Δ_{exper}	Δ_{wav}	Δ_{theor}	Δ_{exper}	Δ_{wav}
0.16	0.987	1.000	0.995	0.997	0.998	0.992
0.32	0.974	0.969	0.980	0.993	0.997	0.997
0.48	0.962	0.950	0.956	0.990	0.991	0.994

Equation (11.13) obtains the form

$$R(i,l) = G(l)H(i,l) + \Omega\{[1 - x_c(l)]p_{3i}(1) - p_{4i}(1) + P_4(i,l)]\}. \tag{11.15}$$

To start off, the uncracked beam is considered. Carrying out computations for $J = 6$ we found $\Omega_1^{(0)} = 12.3627$, $\Omega_2^{(0)} = 485.48$; the exact values are 12.3627 and 485.48, respectively. It follows from here that the accuracy of the wavelet solution is very high already for the level $J = 6$.

Example 11.2: Free vibrations of a cantilever beam with a crack location $x = 0.275$ were considered by Sinha et al. [29]. The relative crack depths were $\beta = 0.16$, $\beta = 0.32$ and $\beta = 0.48$. The solution was obtained by using a FEM model with rotational springs (16 elements with 34 degrees of freedom). Numerical results were checked on the experimental data. Let us compare these data with our wavelet results. The reduction of the frequency of vibrations due to the cracks is estimated by the coefficient

$$\Delta_\nu = \sqrt{\Omega_\nu / \Omega_\nu^{(0)}}, \quad \nu = 1, 2, ... \tag{11.16}$$

Results of the computation are shown in Table 11.2. The symbols Δ_{theor} and Δ_{exper} indicate the theoretical and experimental values taken from Table 11.2 of the Sinha's et al. paper [29], Δ_{wav} corresponds to the wavelet solution. Analysis of these data shows that to be compared with the experiment, our solution is not worse than the Sinha's FEM solution.

The presented method is applicable also for stepped beam. Consider a two-stepped beam for which $\gamma(x) = 1$ for $0 \leq x < \eta$ and $\gamma(x) = \gamma_1$ for $\eta < x \leq 1$. Evaluating the integrals (11.11) we find

$$\begin{aligned} I_{1i}(1) &= \gamma_1 p_{3i}(1) - (1 - \gamma_1) p_{3i}(\eta), \\ I_{2i}(x_l) &= P_4(i,l) + \{-(1 - \gamma_1)[P_i(i,l) - p_{4i}(\eta) - p_{3i}(\eta)(x_l - \eta)]\}U(x_l - \eta), \end{aligned} \tag{11.17}$$

where U is the Heaviside function.

Example 11.3: Free vibrations of a two-stepped beam with two cracks supposed to belong to different steps were analyzed by Zhang et al. [35]; see also Fig. 11.2. The cracks were modelled by springs, the cracks and the step location divided the beam into four sub-beams; the governing equations were solved by the method of transfer matrix. In the following the solution based on the Haar wavelet method is

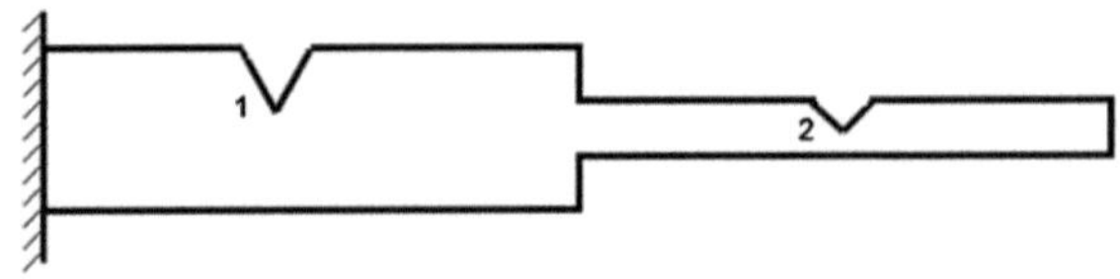

Fig. 11.2 Free vibrations of a two-stepped cantilever beam

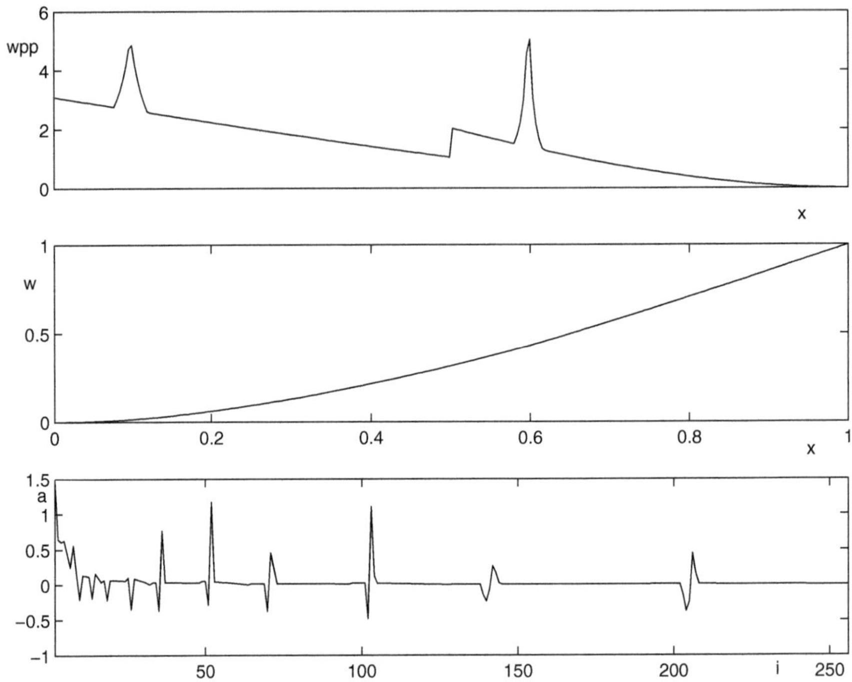

Fig. 11.3 Curvature, deflection and wavelet coefficients for problem 3 from [35]; first mode

presented. Let us assume that the step is located at $x = \eta$, the crack locations are ξ_1, ξ_2 and their depths β_1, β_2.

The cracks are modelled according to (10.15). The cracks effective half-width is taken again in the form $\Lambda = \Gamma h/(0.9L)$; in the case of a stepped beam, the parameter Γ is calculated from the formula $\Gamma = \gamma^3 - (\gamma - \beta)^3$. Computer simulation was carried out for the Cases 3 and 8 of the paper [35]. The results are shown in Figs. 11.3 and 11.4 and in Table 11.3. In this table the symbols Δ_{Zh} correspond to the solutions of paper [35], the symbols Δ_{wav} to our wavelet solution. It follows from Table 11.3 that the accordance of these solutions is very good for the first mode and somewhat worse in the case of the second mode. With reference to Figs. 11.3 and 11.4 we would like to turn attention to the fact that most of the wavelet coefficients are zero, which is typical of the discrete wavelet methods and to a great extent increases the speed of convergence.

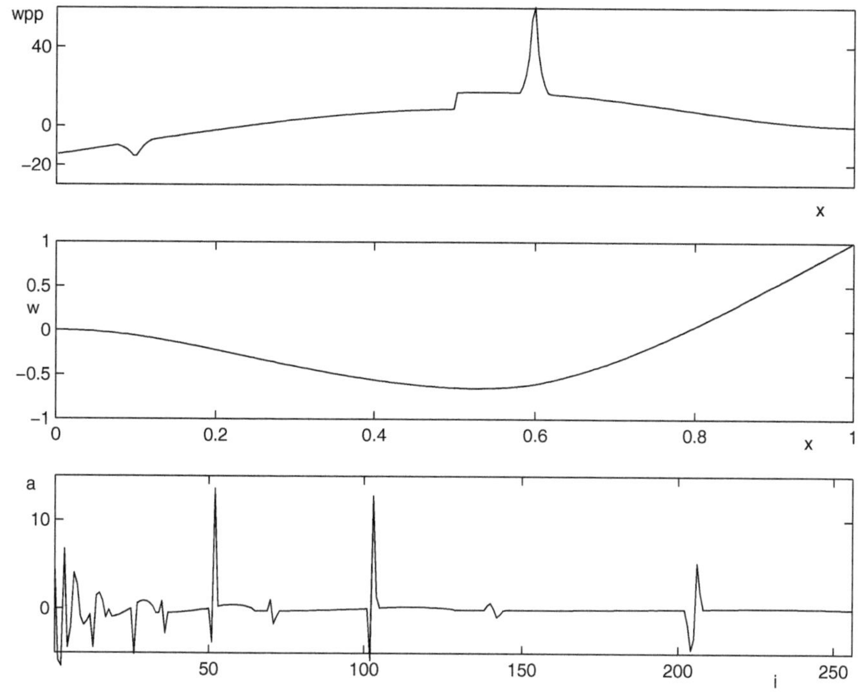

Fig. 11.4 Curvature, deflection and wavelet coefficients for problem 3 from [35]; second mode

Table 11.3 Reduction of eigenfrequencies of cracked beams, comparison of the results with the results by Zhang et al. [35]

Mode	Case 3		Case 8	
	Δ_{Zh}	Δ_{wav}	Δ_{Zh}	Δ_{wav}
1	0.977	0.977	0.947	0.949
2	0.985	0.984	0.948	0.964

The wavelet method is very convenient for treating stepped beams. According to the conventional approach, the beam is divided into sub-beams between the steps; governing equations are solved for each beam separately satisfying the continuity equations at each step. This procedure may turn out to be very troublesome, especially if the number of steps is more than two. The wavelet solution is much more simple since the beam is treated as a whole (not dividing it into sub-beams).

11.4 Forced Vibrations of the Beam

Example 11.4: Vibrations of cracked cantilever beams under a concentrated harmonic load were discussed by Orhan et al. [24]. Single and double V-shaped cracks were considered (Fig. 11.5). For solution, the ANSYS 8.0 program was used and

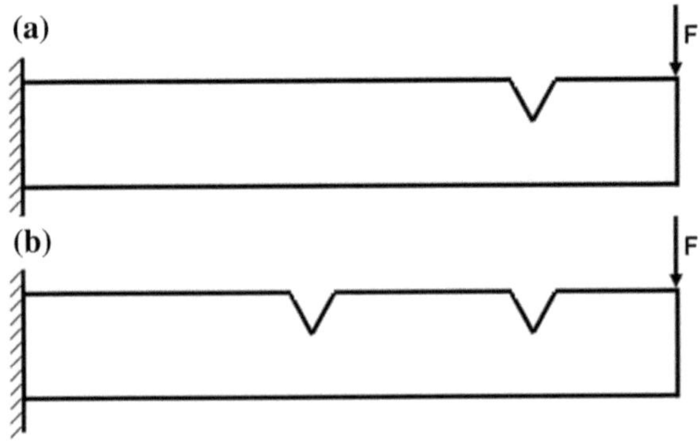

Fig. 11.5 Harmonic vibrations of single and double cracked cantilever beams [24]

the beam was discretized into 1520 elements with 2300 nodes. The beam and material parameters were $L = 0.5\,\text{m}$, $b = 0.2\,\text{m}$, $h = 0.029\,\text{m}$, $E = 206.8\,\text{GPa}$, $\rho = 7780\,\text{kg/m}^3$. Now let us solve the same problem by the wavelet method. For this again (11.4) must be solved. The values of the coefficients in (11.6) are $A = 3.420\text{e-}6$, $B = 1.487\text{e-}5$. Computer simulation was carried out for two cases from [24]: (i) a single crack at $\xi_1 = 0.75$, (ii) two cracks at $\xi_1 = 0.5$, $\xi_2 = 0.75$. In both cases, for the crack depths, the values $d = 8\,\text{mm}$, $d = 20\,\text{mm}$, $d = 24\,\text{mm}$ were taken.

For getting the Haar wavelet solution we specify the angular velocity ω, calculate $R = A\omega^2$ and apply the V-shaped crack model (10.15) by assuming $\Lambda = \Gamma h/(0.9L)$. The matrix $R(i, l)$ is calculated according to (11.13); the wavelet coefficients are evaluated by solving the system (11.12). This course of solution is carried out for different values of ω.

For analyzing the obtained numerical data we calculate the end deflection $W = |w(1)|$ as a function on the forcing frequency $f_r = \omega/2\pi$. For an undamaged beam, this result is plotted in Fig. 11.6. It follows from this figure that here, from a background of low intensity, a sharp peak arises. This is a phenomenon of resonance. The critical value of the forcing frequency is $\omega_r = 605.18$ (this is an exact value). The Haar wavelet approach for $J = 7$ gives $\omega_r = 605.19$ or $f_r = \omega_r/2\pi = 96.3$, which practically coincides with the exact value.

In the case of cracked beams, the resonance frequencies decrease. In the case of a single crack this effect is quite small (the same conclusion is made in [24]). For beams with two cracks the computer simulation gave: $f_r = 95.2(96)$ for $d = 8\,\text{mm}$, $f_r = 88.3(88)$ for $d = 20\,\text{mm}$ and $f_r = 77.0(80)$ for $d = 24\,\text{mm}$ (in parentheses are the values of the dominant peaks from [24]). Although the peak values of both solutions are quite close there is nevertheless an essential difference between them.

In the case of the Haar wavelet solution, the curves $W = W(f_r)$ have sharp resonance peaks, but for the FEM solution, a broad specter of forcing frequencies appears. This can be seen from Figs. 11.7 and 11.8 which are taken from the paper by Orhan [24]. Our solution seems to be more logical, since in the case of non-viscous systems, sharp resonance peaks must appear. In our mind, by applying FEM, some

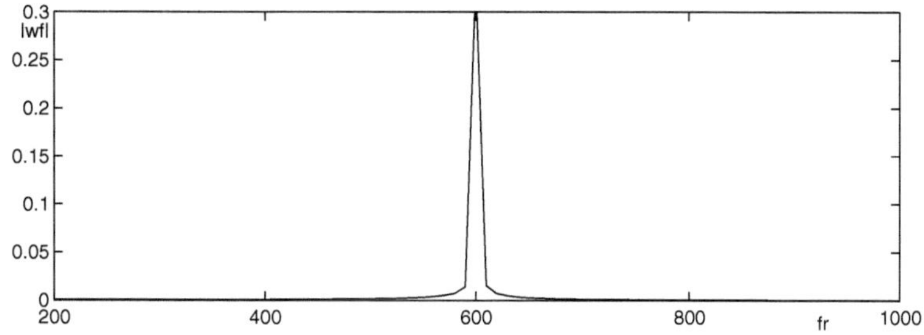

Fig. 11.6 Maximal deflection of the uncracked beam versus forcing frequency

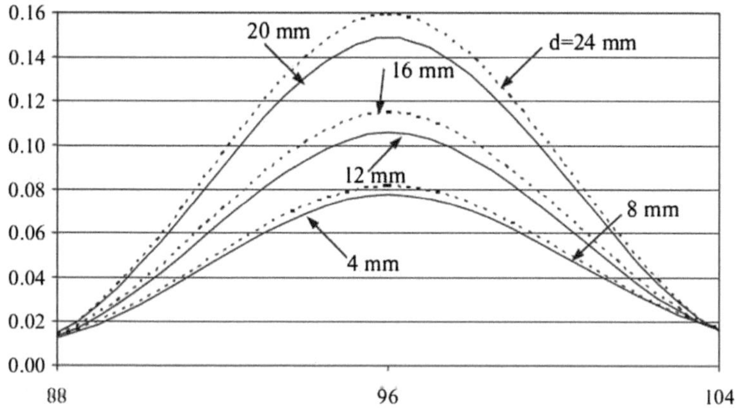

Fig. 11.7 The curves $W = W(f_r)$ fot the beam with a single crack; number denotes the crack depth d (this is a off-print of the Fig. 11.9 from [24])

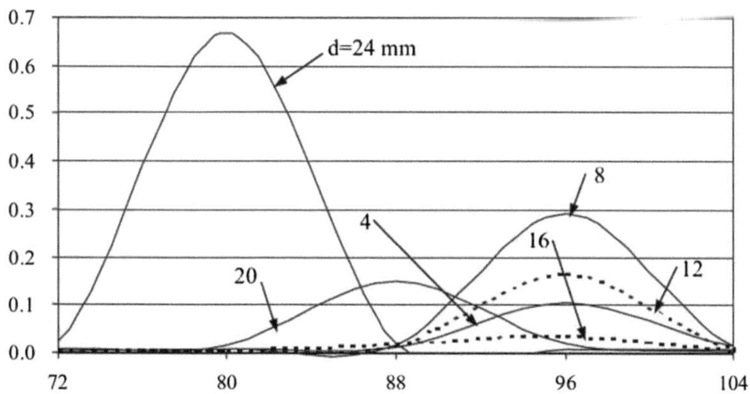

Fig. 11.8 The curves $W = W(f_r)$ fot the beam with two cracks; number denotes the crack depth d (this is a off-print of the Fig. 16 from [24])

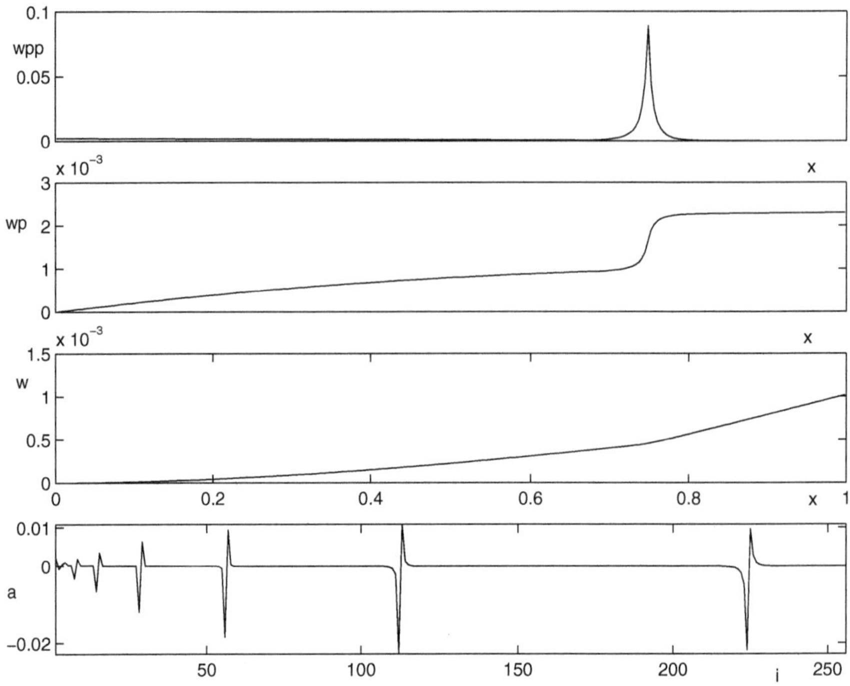

Fig. 11.9 Forced vibrations of the beam; single crack, precritical stage

artificial or hidden viscosity was introduced and due to it the signal was smeared out to some region.

For some cases, the curvature, slope and deflection versus beam coordinate and also the wavelet coefficients were calculated. Single and double cracks with the depth $d = 24$ mm were assumed. The cases $\omega_r = 300$ (precritical stage) and $\omega_r = 800$ (postcritical stage) were considered. The results are plotted in Figs. 11.9, 11.10, 11.11 and 11.12.

11.5 Related Papers

For analyzing bending and vibrations of elastic structures, various wavelet-based solutions are proposed. Luo et al. [20] applied for vibration modelling the fast Gaussian wavelets. Newland [23] and Neild et al. [22] used for structural vibration analysis the harmonic wavelets. Elastic problems were examined with the aid of Daubechies wavelets by Díaz et al. [5], Liu et al. [19] and Ma et al. [21]. Lardies and Gouttebroze [17] applied the Morlet wavelet for identification of modal parameters. B-spline wavelets were applied by Xiang et al. [32, 33]; Chen and Wu [2, 3] applied

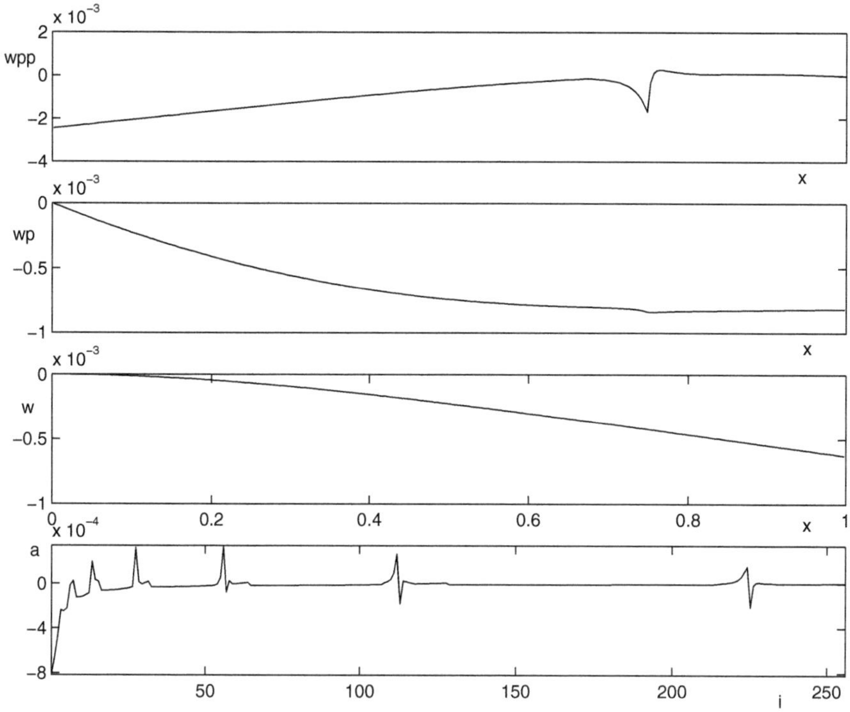

Fig. 11.10 Forced vibrations of the beam; single crack, postcritical stage

these wavelets for membrane vibration analysis. Zhou and Zhou [37] proposed a modified wavelet approximation for calculating deflections of beams and square thin plates.

Bending waves in beams on viscoelastic random foundation were analyzed using wavelet techniques by Koziol and Hryniewicz [15]. Optimization of wavelet transform for impulsive vibration signals was considered by Schukin et al. [28]. Vibratory analysis in two dimensional layers due to a moving load was executed by Koziol et al. [16].

In several papers, wavelet-based solutions are combined with the finite element analysis. Construction of wave finite elements and their applications was discussed by Chen et al. [4]. Daubechies wavelets were applied by Diaz et al. [5] and Ma et al. [21]. Applications of the spline wavelet finite element method were investigated by Han et al. [10, 11] and also by Zhang et al. [36]. Finite element analysis of beam structures based on trigonometric wavelets was presented by He and Ren [13].

There are numerous papers in which the wavelet analysis is applied for crack detection or damping estimation in elastic structures. A detailed analysis of these papers is beyond the scope of this book and we confine ourselves only to some

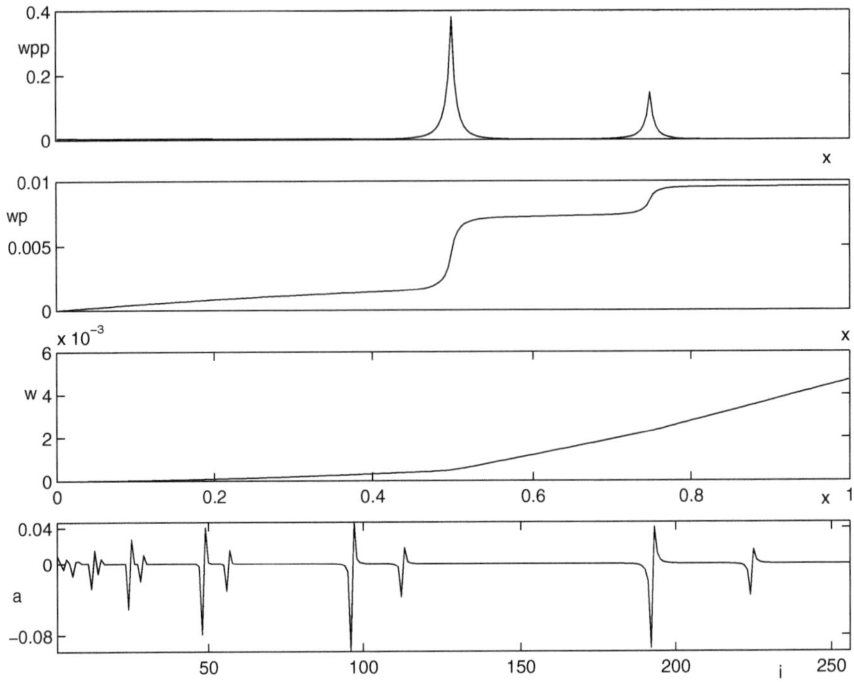

Fig. 11.11 Forced vibrations of the beam; two cracks, precritical stage

samples. Crack detection was discussed in [1, 6, 8, 25, 26]. Damage detection and analysis was carried out in [1, 7, 9, 27].

Our special interest is focused on papers in which Haar wavelet method is applied for analysis of elastic structures. In addition to our paper [18] in which free and forced vibrations of cracked beams were discussed, we note here the paper by Hariharan [12]. In this paper, the bending of Euler-Bernoulli beams on the Winkler foundation are solved. Wang and Deng [31] investigated damage detection in Euler-Bernoulli beams

$$EIw^{IV} + \rho A\ddot{w} = q. \tag{11.18}$$

Most of the analysis was carried out using the Haar wavelets. Different problems (a cantilever beam under impact loading, plate with through-thickness crack) were solved.

Nondestructive damage evaluation of plates using the two-dimensional Haar wavelets was investigated by Kim et al. [14].

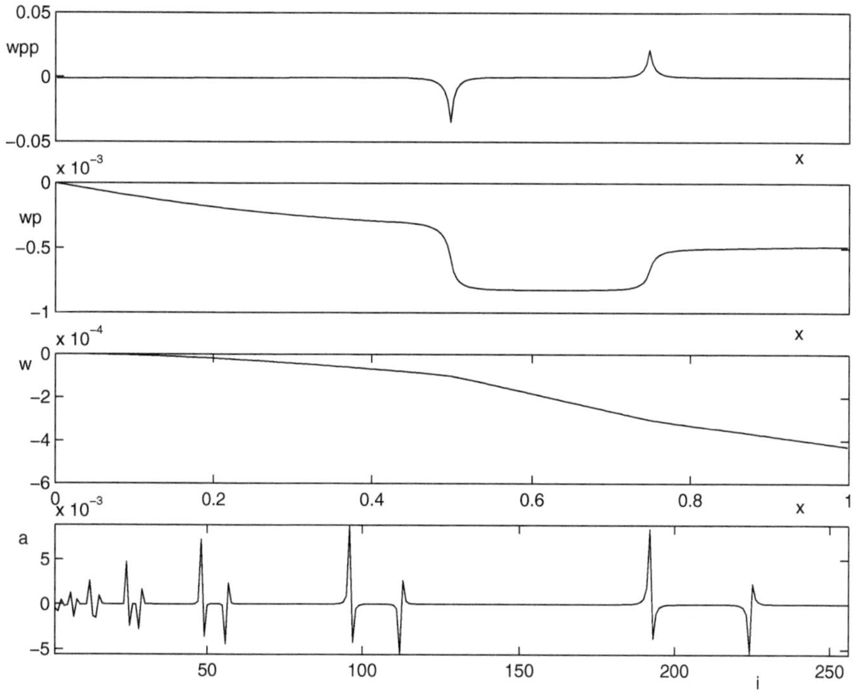

Fig. 11.12 Forced vibrations of the beam; two cracks, postcritical stage

Haar-type orthogonal matrices, which are rarely utilized in practice, were applied for mechanic signal analysis by Xiang et al. [34].

References

1. Chang, C., Chen, L.: Detection of the location and size of cracks in the multiple cracked beam by spatial wavelet based approach. Mech. Syst. Sig. Proc. **19**, 139–155 (2005)
2. Chen, W., Wu, C.: Adaptable spline element formembrane vibration analysis. Int. J. Numer. Meth. Eng. **39**, 2457–2476 (1996)
3. Chen, W., Wu, C.: Extension of spline wavelets element method to membrane vibration analysis. Comput. Mech. **18**, 46–54 (1996)
4. Chen, X., Yang, S., Ma, J., He, Z.: The construction of wavelet finite element and its application. Finite Elem. Anal. Des. **40**, 541–554 (2004)
5. Díaz, L.A., Martín, M., Vampa, V.: Daubechies wavelet beam and plate finite elements. Finite Elem. Anal. Des. **45**, 200–209 (2009)
6. Douka, E., Loutridis, S., Trochidis, A.: Crack identification in beams using wavelet analysis. Int. J. Solids Struct. **40**, 3557–3569 (2003)
7. Fan, W., Qiao, P.: A 2-D continuous wavelet transform of mode shape data for damage detection of plate structures. Int. J. Solids Struct. **46**, 4379–4395 (2009)
8. Gentile, A., Messina, A.: On the continuous wavelet transforms applied to discrete vibrational data for detecting open cracks in damaged beams. Int. J. Solids Struct. **40**, 295–315 (2003)

9. Grabowska, J., Palacz, M., Krawczuk, M.: Damage identification by wavelet analysis. Mech. Syst. Sig. Proc. **22**, 1623–1635 (2008)
10. Han, J., Ren, W., Huang, Y.: A spline wavelet finite-element method in structural mechanics. Int. J. Numer. Meth. Eng. **66**, 166–190 (2006)
11. Han, J., Ren, W., Huang, Y.: A spline wavelet finite element formulation of thin plate bending. Engineering with Computers **25**, 319–326 (2009)
12. Hariharan, G.: Solving finite length beam equation by the Haar wavelet method. Int. J. Comput. Appl. **9**, 27–34 (2010)
13. He, W., Ren, W.: Finite element analysis of beam structures based on trigonometric wavelet. Finite Elem. Anal. Des. **51**, 59–66 (2012)
14. Kim, B., Kim, H., Park, T.: Nondestructive damage evaluation of plates using the multi-resolution analysis of two-dimensional Haar wavelet. J. Sound Vib. **292**, 82–104 (2006)
15. Koziol, P., Hryniewicz, Z.: Analysis of bending waves in beam on viscoelastic random foundation using wavelet technique. Int. J. Solids Struct. **43**, 6965–6977 (2006)
16. Koziol, P., Mares, C., Esat, I.: Wavelet approach to vibratory analysis of surface due to a load moving in the layer. Int. J. Solids Struct. **45**, 2140–2159 (2008)
17. Lardies, J., Gouttebroze, S.: Identification of modal parameters using the wavelet transform. Int. J. Mech. Sci. **44**, 2263–2283 (2002)
18. Lepik, Ü.: Exploring vibrations of cracked beams by the Haar wavelet method. Est. J. Eng. **18**, 58–75 (2012)
19. Liu, Y., Qin, F., Liu, Y., Cen, Z.: A Daubechies wavelet-based method for elastic problems. Eng. Anal. Boundary Elem. **34**, 114–121 (2010)
20. Luo, G., Osypiw, D., Irle, M.: Vibration modelling with fast Gaussian wavelet algorithm. Adv. Eng. Softw. **33**, 191–197 (2002)
21. Ma, J., Xue, J., Yang, S., He, Z.: A study of the construction and application of a Daubechies wavelet-based beam element. Finite Elem. Anal. Des. **39**, 965–975 (2003)
22. Neild, S., McFadden, P., Williams, M.: A review of time-frequency methods for structural vibration analysis. Eng. Struct. **25**, 713–728 (2003)
23. Newland, D.: An Introduction to Random Vibrations, Spectral and Wavelet Analysis. Longman, London (1993)
24. Orhan, S.: Analysis of free and forced vibration of a cracked cantilever beam. NDT&E Int. **40**, 443–450 (2007)
25. Quek, S., Wang, Q., Zhang, L., Ang, K.: Sensitivity analysis of crack detection in beams by the wavelet technique. Int. J. Mech. Sci. **43**, 2899–2910 (2001)
26. Rizos, P., Aspragathos, N., Dimarogonas, A.: Identification of crack location and magnitude in a cantilever beam from the vibration modes. J. Sound Vib. **138**, 381–388 (1990)
27. Rucka, M., Wilde, K.: Application of continuous wavelet transform in vibration based damage detection method for beams and plates. J. Sound Vib. **297**, 536–550 (2006)
28. Schukin, E., Zamaraev, R., Schukin, L.: The optimisation of wavelet transform for the impulse analysis in vibration signals. Mech. Syst Sign. Proces. **18**, 1315–1333 (2004)
29. Sinha, J., Friswell, M., Edwards, S.: Simplified models for the location of cracks in beam structures using measured vibration data. J. Sound Vibr. **251**, 13–38 (2002)
30. Skrinar, M.: Elastic beam finite element with an arbitrary number of transverse cracks. Finite Elem. Anal. Des. **45**, 181–189 (2009)
31. Wang, Q., Deng, X.: Damage detection with spatial wavelets. Int. J. Solids Struct. **36**, 3443–3468 (1999)
32. Xiang, J., Chen, X., He, Z., Zhang, Y.: A new wavelet-based thin plate element using B-spline wavelet on the interval and its applications. Comput. Mech. **41**, 243–255 (2008)
33. Xiang, J., Chen, X., Yang, L., He, Z.: A class of wavelet-based flat elements using B-spline wavelet on the interval and its applications. Comput. Model. Eng. Sci. **23**, 1–12 (2008)
34. Xiang, X., Zhou, J., Yang, J., Liu, L., An, X., Li, C.: Mechanic signal analysis based on the Haar-type orthogonal matrix. Expert Syst. Appl. **36**, 9674–9677 (2009)

35. Zhang, W., Wang, Z., Ma, H.: Crack identification in stepped cantilever beam combining wavelet analysis with transform matrix. Acta Mech. Solida Sin. **22**, 360–368 (2009)
36. Zhang, X., Chen, X., Wang, X., He, Z.: Multivariable finite elements based on B-spline waveleton the interval for thin plate static and vibration analysis. Finite Elem. Anal. Des. **46**, 416–427 (2010)
37. Zhou, Y., Zhou, J.: A modified wavelet approximation of deflections for solving PDEs of beams and square thin plates. Finite Elem. Anal. Des. **44**, 773–783 (2008)

Chapter 12
Free Vibrations on Non-uniform and Axially Functionally Graded Euler-Bernoulli Beams

12.1 Governing Equations

In the present chapter, the vibration analysis of non-uniform and functionally graded (FG) Euler-Bernoulli beams is considered [9]. Functionally graded materials have attracted a great deal of attention from researchers and engineers due to their unique properties, such as thermal resistance, high toughness, and low density. In the case of functionally graded beams, the gradient may vary along the cross-section or/and along the axial direction.

Let us consider a vibrating FG Euler-Bernoulli beam of length L. It is assumed that the beam has a variable cross-section and the material properties vary along the beam axis coordinate $\tilde{x}$. Introducing the quantities

$$x = \frac{\tilde{x}}{L}, \; k^4 = \frac{\rho_0 A_0 \omega^2 L^4}{E_0 I_0}, \tag{12.1}$$

the equation of motion for transverse vibrations is given by

$$\frac{d^2}{dx^2}[E(x)I(x)\frac{d^2 W(x)}{dx^2}] - k^4 m(x)W(x) = 0, \; x \in [0, 1] \tag{12.2}$$

where $W(x)$ is the transverse deflection, $m(x) = \rho(x)A(x)$ is the mass at position x, $E(x)I(x) = D(x)$ is the bending stiffness; $\rho(x)$ is the mass density of the beam material, $E(x)$ is the Young's modulus, $A(x)$ is the cross-section area and $I(x)$ is the moment of inertia at x. The quantities $\rho_0, \; A_0, \; E_0, \; I_0$ denote the values of $\rho, \; A, \; E, \; I$ at $x = 0$, respectively. From (12.2) yields that

Ü. Lepik and H. Hein, *Haar Wavelets*, Mathematical Engineering,
DOI: 10.1007/978-3-319-04295-4_12,

$$\frac{d^4W(x)}{dx^4}E(x)I(x) + 2\frac{d^3W(x)}{dx^3}\left[\frac{dE(x)}{dx}I(x) + \frac{dI(x)}{dx}E(x)\right]$$
$$+\frac{d^2W(x)}{dx^2}\left[\frac{d^2E(x)}{dx^2}I(x) + 2\frac{dE(x)}{dx}\frac{dI(x)}{dx} + \frac{d^2I(x)}{dx^2}E(x)\right]$$
$$- k^4W(x)\rho(x)A(x) = 0,\ x \in [0,\ 1]. \tag{12.3}$$

The Eq. (12.1) is fourth order ODE with variable coefficients for which the analytic solution in a general case is not available. According to the Haar wavelet method, the solution of (12.1) is sought in the form:

$$W^{IV}(x) = \sum_{i=1}^{2M} a_i h_i(x). \tag{12.4}$$

Integrating (12.4) four times we obtain

$$\begin{aligned}
W'''(x) &= \textstyle\sum_{i=1}^{2M} a_i p_{1,i}(x) + W'''(0),\\
W''(x) &= \textstyle\sum_{i=1}^{2M} a_i p_{2,i}(x) + W'''(0)x + W''(0),\\
W'(x) &= \textstyle\sum_{i=1}^{2M} a_i p_{3,i}(x) + \frac{1}{2}W'''(0)x^2 + W''(0)x + W'(0),\\
W(x) &= \textstyle\sum_{i=1}^{2M} a_i p_{4,i}(x) + \frac{1}{6}W'''(0)x^3 + \frac{1}{2}W''(0)x^2 + W'(0)x + W(0).
\end{aligned} \tag{12.5}$$

In (12.5), $p_{\alpha i},\ \alpha = 1, \ldots, 4$ are calculated from (2.6) to (2.7); the quantities $W(0),\ W'(0),\ W''(0),\ W'''(0)$ can be evaluated from the boundary conditions. Let us consider the following boundary conditions:

(i) Cantilever beam (CF)

In the case of a cantilever beam the system derived from (12.5) can be presented as

$$\begin{aligned}
&\textstyle\sum_{i=1}^{2M} a_i p_{1,i}(1) + W'''(0) = 0,\\
&\textstyle\sum_{i=1}^{2M} a_i p_{2,i}(1) + W'''(0) + W''(0) = 0,
\end{aligned} \tag{12.6}$$

from which $W''(0)$ and $W'''(0)$ are obtained

$$\begin{aligned}
W''(0) &= \textstyle\sum_{i=1}^{2M} a_i q_{1,i},\\
W'''(0) &= -\textstyle\sum_{i=1}^{2M} a_i p_{1,i}(1).
\end{aligned} \tag{12.7}$$

In (12.7) $q_{1,i} = p_{1,i}(1) - p_{2,i}(1)$. Substituting (12.7) into (12.5) the mode shape W obtains the following form:

$$W(x) = \sum_{i=1}^{2M} a_i \left[p_{4,i}(x) - \frac{1}{6}p_{1,i}(1)x^3 + \frac{1}{2}q_{1,i}x^2\right]. \tag{12.8}$$

(ii) Pinned-pinned beam (PP)

In the case of pinned of both ends beam from (12.5) yields

$$\begin{aligned}&\sum_{i=1}^{2M} a_i p_{2,i}(1) + W'''(0) = 0,\\&\sum_{i=1}^{2M} a_i p_{4,i}(1) + \tfrac{1}{6}W'''(0) + W'(0) = 0,\end{aligned} \tag{12.9}$$

from which we get

$$\begin{aligned}W'''(0) &= -\sum_{i=1}^{2M} a_i p_{2,i}(1),\\W'(0) &= \frac{1}{6}\sum_{i=1}^{2M} a_i q_{2,i},\\W(x) &= \sum_{i=1}^{2M} a_i \left[p_{4,i}(x) - \frac{1}{6}p_{2,i}(1)x^3 + \frac{1}{6}q_{2,i}x\right],\end{aligned} \tag{12.10}$$

where $q_{2,i} = p_{2,i}(1) - 6p_{4,i}(1)$.

(iii) Clamped-clamped beam (CC)

Satisfying the boundary conditions for the clamped-clamped beam one could get

$$\begin{aligned}&W'''(0) = 6\sum_{i=1}^{2M} a_i q_{3,i},\\&W''(0) = \sum_{i=1}^{2M} a_i q_{4,i},\\&W(x) = \sum_{i=1}^{2M} a_i \left[p_{4,i}(x) + q_{3,i}x^3 + \tfrac{1}{2}q_{4,i}x^2\right],\end{aligned} \tag{12.11}$$

where $q_{3,i} = 2p_{4,i}(1) - p_{3,i}(1)$ and $q_{4,i} = 2p_{3,i}(1) - 6p_{4,i}(1)$.

(iv) Clamped-pinned (CP) beam

In this case from (12.5) yields

$$\begin{aligned}W'''(0) &= \sum_{i=1}^{2M} a_i q_{5,i},\\W''(0) &= \sum_{i=1}^{2M} a_i q_{6,i},\\W(x) &= \sum_{i=1}^{2M} a_i \left[p_{4,i}(x) + \frac{1}{6}q_{5,i}x^3 + \frac{1}{2}q_{6,i}x^2\right],\end{aligned} \tag{12.12}$$

where $q_{5,i} = 3p_{4,i}(1) - \frac{3}{2}p_{2,i}(1)$ and $q_{6,i} = -q_{5,i} - p_{2,i}(1)$. For the illustration of the method, consider a case of the cantilever beam. Next, the notation is introduced as follows:

$$a(:)H(:,l) = \sum_{i=1}^{2M} a_i h_i(x_l), \tag{12.13}$$

where H is the Haar matrix with elements $H(i,l) = h_i(x_l)$. Substituting (12.5), (12.7) and (12.8) into (12.1), taking into account (12.13) and discretizating the results by taking $x \to x_l$, the governing system is derived

$$
\begin{aligned}
& a(:)\Big\{H(:,l)U_1(l) + 2\left[P^{(1)}(:,l) - P^{(1)}(:,1)E(l)\right]U_2(l) \\
& + \left[P^{(2)}(:,l) - P^{(1)}(:,1)x_l E(l) + q_1(:)x_l^2 E(l)\right]U_3(l) \\
& -k^4\left[P^{(4)}(:,l) - \frac{1}{6}P^{(1)}(:,1)x_l^3 E(l) + \frac{1}{2}q_1(:)x_l^2 E(l)\right]U_4(l)\Big\}, \\
& = 1,\ldots,2M,
\end{aligned}
\tag{12.14}
$$

where P is the matrix with elements $P(i,l) = p_i(x_l)$ and

$$
\begin{aligned}
U_1(l) &= E(x_l)I(x_l), \\
U_2(l) &= E'(x_l)I(x_l) + E(x_l)I'(x_l), \\
U_3(l) &= E''(x_l)I(x_l) + 2E'(x_l)I'(x_l) + E(x_l)I''(x_l), \\
U_4(l) &= \rho(x_l)A(x_l)
\end{aligned}
\tag{12.15}
$$

and $E(l)$ is a unit row vector. The system (12.14) is linear and homogeneous with regard to a_i and contains the frequency parameter ω. For deriving a nontrivial solution, the determinant of system (12.14) must be zero. According to this requirement, the values for ω are evaluated.

12.2 Beams with Non-uniform Cross-Section and Elastic End Constraints

Example 12.1: Consider a wedge beam with a rectangular cross-section and clamped-free ends. It is assumed that the breadth and the height of the beam are described by the formulae:

$$
h(x) = h_0\left[1 + (\alpha_h - 1)x\right], \quad b(x) = b_0\left[1 + (\alpha_b - 1)x\right], \tag{12.16}
$$

where α_b stands for the ratio between the breadths at the beginning and at the end of the beam; α_h is the ratio between the heights respectively. In Table 12.1, the first dimensionless natural frequencies (DNF) k^2 for the wedge beam with $\alpha_b = 1$, $\alpha_h = \alpha$ and the cone beam with $\alpha_b = \alpha_h = \alpha$ are presented. The results are compared with the results calculated by Hsu [10] which are given in columns two and four. According to the results, the Haar wavelets approach works accurately with insignificant errors in the cases of simple wedge beam models.

Next, consider a non-uniform beam with elastic end constants. The boundary conditions in the presence of translational and rotational spring constants at $x = 0$ are presented as [10]:

Table 12.1 The first DNF k^2 for the wedge and cone cantilever

α	Wedge beam		Cone beam	
	[10]	Present	[10]	Present
0.1	4.63074	4.6305	7.20500	7.2055
0.4	3.93428	3.9343	5.00906	5.0088
0.6	3.73708	3.7373	4.31879	4.3189
0.7	3.66675	3.6670	4.06694	4.0671
0.9	3.55870	3.5589	3.67371	3.6739

Table 12.2 The first two DNF k_n^2 for a beam with translational constraints and $k_{rr} = k_{rl} = 0$

$k_{tl} = k_{tr}$	n = 1		n = 2	
	[10]	Present	[10]	Present
0.001	0.21656	0.2166	0.31795	0.3180
0.01	0.38510	0.3851	0.5639	0.5654
0.1	0.68462	0.6846	1.00528	1.0053
1	1.21404	1.2140	1.78509	1.7851
10	2.10096	2.1009	3.13023	3.1303
100	3.07241	3.0723	5.06670	5.0668
1000	3.37553	3.3754	6.56963	6.5697

$$\frac{d}{dx}[I(x)\frac{d^2w(x)}{dx^2}] + k_{tl}w(x) = 0, \tag{12.17}$$

$$I(x)\frac{d^2w(x)}{dx^2} - k_{rl}\frac{dw(x)}{dx} = 0. \tag{12.18}$$

In (12.17) and (12.18), the non-dimensional translational and rotational spring coefficients k_t and k_r are

$$k_{tl} = \frac{K_{TL}L^3}{EI_0}, \; k_{rl} = \frac{K_{RL}L}{EI_0}, \tag{12.19}$$

where K_{TL} and K_{RL} correspond to the translational and rotational spring constants at the flexible left end, respectively. The boundary conditions on the right end $x = 1$ are defined analogically. The first two natural frequencies k_n^2, $n = 1, 2$ with $\alpha_b = \alpha_h = 1.4$, fixed rotational spring constraints $k_{rr} = k_{rl} = 0$ and varying translational spring constraints are presented in Table 12.2. The results are compared with Hsu's [10] calculations.

The Haar wavelet approach was also applied to the beams with varying translational and fixed rotational end constraints ($k_{rr} = 0.5$, $k_{rl} = 1$) with $\alpha_b = 2$ and $\alpha_h = 1$ (Table 12.3).

The study of the frequencies proves that there is slight dependence between the rotational constraints and the cross-section area. The influence can be observed with the growth of the translational constraint value.

Now consider a truncated at α tapered beam [26] with parabolic-taper width $b(x) = b_0\sqrt{x}$ and linear-taper height $h(x) = h_0 x$. The first three DNF for boundary

Table 12.3 The first DNF k^2 for the beam with varying translational and fixed rotational end constraints

$k_{tl} = k_{tr}$	$k_{rr} = 0.5$, $k_{rl} = 1$	$k_{rr} = 1$, $k_{rl} = 0.5$
0.001	0.2115	0.2115
0.01	0.3760	0.3760
0.1	0.6682	0.6684
1	1.1803	1.1828
10	2.0210	2.0345
100	2.9628	2.9852

Table 12.4 The first three DNF k_n^2 for the beam with parabolic-taper width and linear-taper height

α	n = 1		n = 2		n = 3	
	Present	[26]	Present	[26]	Present	[26]
0.1	5.8383	5.8382	16.6901	16.696	34.8190	34.854
0.3	4.7574	4.7577	17.5988	17.600	41.6669	41.660
0.5	4.2101	4.2100	18.9250	18.922	47.9306	47.907
0.8	3.7303	3.7301	20.8437	20.838	56.4426	56.453

Table 12.5 The first DNF of the cantilever with parabolic thickness (rectangular and circular cross-section) versus the dimensionless coordinate of the fixed end

x_1	Cross-section			
	Rectangular		Circular	
	[3]	Present	[3]	Present
−0.7	1.050	1.0474	1.011	1.0028
−0.5	1.936	1.9363	2.238	2.2357
−0.3	3.070	3.0704	3.940	3.9394
0.0	5.576	5.5774	7.886	7.8857
0.3	10.17	10.1784	15.31	15.3134
0.5	16.27	16.2805	25.26	25.2880
0.7	30.47	30.4827	48.49	48.4945

conditions (on the left end $k_{rl} = k_{tl} = 0$ and on the right end $k_{rr} \to \infty$, $k_{tr} \to \infty$) with $J = 5$ are presented in Table 12.4. The calculated results correspond well with the previous work [26]. Next, examine the Haar wavelet approach on the cantilever with parabolic thickness versus the dimensionless coordinate of the fixed end ($h = h_0(1 - x^2)$) and the cantilever with circular cross-section and parabolic thickness. In this case α is the dimensionless coordinate of the fixed end. The reference cross-section area and moment of inertia, A_0 and I_0 are considered at the longitudinal coordinate $x = 0$. The results of the calculations and comparison with results in [3] are provided in Table 12.5. The calculated results correspond with the previous research.

Table 12.6 The first DNF for beams the with variable flexural rigidity, mass density and elastic end constraints ($k_{rr} = k_{tr} = 1$)

$\alpha = \beta$	$k_{rl} = k_{tl}$				
	1.0	2.0	3.0	4.0	5.0
−0.2	1.4408	2.0732	2.5756	3.0206	3.4420
−0.15	1.4120	2.0309	2.5191	2.9473	3.3462
0.15	1.2950	1.8632	2.3030	2.6794	3.0176
0.2	1.2818	1.8446	2.2795	2.6510	2.9842

12.3 Non-homogeneous Beams with Variable Flexural Rigidity, Mass Density and Elastic End Constraints

Example 12.2: Let us consider the cases of axially FG beams with unusual boundary conditions. Assume that the flexural rigidity and mass density of the beam vary in the following form:

$$D(x) = D_0\left[1 + \alpha cos(\pi x)\right],\ \rho(x) = \rho_0\left[1 + \beta cos(\pi x)\right], \tag{12.20}$$

where $|\alpha| < 1$ and $|\beta| < 1$ are parameters. The conditions ensure that $D(x)$ and $\rho(x)$ are positive. The ends of the beam are fixed by elastic spring supports. The boundary conditions in the presence of rotational and translational spring constants are given by Eqs. (12.17)–(12.19). In Table 12.6, the first natural frequencies are presented for the case of fixed translational and rotational spring constants at the right end ($k_{rr} = k_{tr} = 1$) and variable constants at the left end.

12.4 Non-homogeneous Beams with Variable Flexural Rigidity, Mass Density, Elastic End Constraints and Intermediate Rigid Support

Example 12.3: Finally, the present method can also be applied for beams with additional intermediate constraints. Let us consider an axially FG beam with variable flexural rigidity and mass density varying according to the Eq. (12.20). The beam has rotational and translational flexible ends and an additional rigid support at $x = \gamma$. In the presence of the rigid support, the condition is

$$W(\gamma) = 0. \tag{12.21}$$

According to (12.12), the Eq. (12.21) takes the form:

$$W(\gamma) = \sum_{i=1}^{2M} a_i \left[p_{4,i}(\gamma) + \frac{1}{6} q_{5,i}\gamma^3 + \frac{1}{2} q_{6,i}\gamma^2\right]. \tag{12.22}$$

Table 12.7 The first DNF for beams with variable flexural rigidity, elasticend constraints and intermediate rigid support ($k_{rr} = k_{tr} = k_{rl} = k_{tl} = 1$)

$\alpha = \beta$	γ				
	0.1	0.3	0.5	0.7	0.9
−0.2	1.8902	2.8150	4.0124	3.9385	3.2176
−0.15	1.9120	2.8620	4.0277	3.8784	3.1657
0.15	2.0578	3.1553	4.0527	3.5405	2.8874
0.2	2.0851	3.2077	4.0475	3.4879	2.8455

In Table 12.7, the calculated first natural frequencies for the fixed translational and rotational spring constants $k_{rr} = k_{tr} = k_{rl} = k_{tl} = 1$, varying $\alpha = \beta$ and different values of γ are presented. It can be seen in Table 12.7 that in the case of symmetric boundary conditions, the natural frequencies at the right end and left end are not the same. This could be explained by non-symmetric FG material distribution in the beam.

12.5 Related Papers

The only work known to authors in which FG structures are studied with the aid of the Haar wavelets has been published by Chun and Zheng [5]. They considered the simply-supported rectangular plates with an arbitrary distribution of material properties in a transverse direction.

In most of the works which consider the FG beams, it is assumed that the gradient varies along the cross-section, e.g. [1, 4, 6–8, 11, 14, 15, 18, 21, 23, 25]. Here we concentrate on the papers in which the axially FG beams and/or axially nonhomogeneous beams are considered.

The beams with varying cross-section were investigated in [3, 10, 26]. For axially functionally graded (AFG) beams the governing equation could be transformed to the ODE with variable coefficients. So far, few analytical solutions are found for arbitrary gradient change due to the difficulty of mathematical treatment of the problems. Huang and Li [12] transformed the governing equation to Fredholm integral equations. Natural frequencies can be determined by requiring that the resulting Fredholm integral equation has a non-trivial solution. The method is illustrated by calculating the frequencies for tapered beams of linearly variable width and graded beams of special polynomial non-homogeneity. In [11, 13] Huang and Luo proposed a simple method to determine the buckling load of inhomogeneous AFG beams. By expanding the mode shape into a power series, the governing equation was transformed a the system of linear algebraic equations. The flexural rigidity was considered as polynomial or exponential. Buckling of AFG beams was considered in [11, 22]. In [17] the differential transformation method and differential quadrature element method were applied for free bending vibration of rotating AFG Euler Bernoulli tapered beams. Şimşek et al. [20] investigated AFG beams under action of a moving load and the dynamic responses of AFG beams were obtained with the aid of the Newmark method. Li et al. [16] studied AFG beams where the bending stiffness and mass distribution was considered according to the exponential law. The characteristic equations are derived in closed form. The gradient's and support conditions' influence on the frequency spectrum was also investigated. Shahba and Rajasekaran [19] introduced the differential transform element method and the differential quadrature element method of the lowest-order to solve the governing differential equation. The free vibration and stability of linearly tapered AFG, where the mass density and elastic modulus vary according to the power law, were investigated. Wu et al. [24] studied the case where the axially graded structure has a prescribed polynomial second mode. The proposed method is feasible when the inertial coefficient, flexural stiffness and the mode shape are polynomial functions. Hsu et al. [10] considered free vibration of non-uniform Euler-Bernoulli beams with general elastically end constraints using the Adomian modified decomposition method. The finite element method was applied for a functionally graded beam with material graduation in axially or transversally through the thickness based on the power law by Alshorbagy et al. [2].

References

1. Allahverdizadeh, A., Mahjoob, M., Eshraghi, I., Nasrollahzadeh, N.: On the vibration behavior of functionally graded electrorheological sandwich beams. Int. J. Mech. Sci. **70**, 130–139 (2013)
2. Alshorbagy, A., Eltaher, M., Mahmoud, F.: Free vibration characteristics of a functionally graded beam by finite element method. Appl. Math. Model. **35**, 412–425 (2011)
3. Caruntu, D.: Dynamic modal characteristics of transverse vibrations of cantilevers of parabolic thickness. Mech. Res. Commun. **36**, 391–404 (2009)
4. Chakraborty, A., Gopalakrishnan, S., Reddy, J.: A new beam finite element for the analysis of functionally graded materials. Int. J. Mech. Sci. **45**, 519–539 (2003)
5. Chun, Z., Zheng, Z.: Three-dimensional analysis of functionally graded plate based on the Haar wavelet method. Acta Mech. Solida Sin. **20**, 95–102 (2007)
6. Dai, H., Zheng, H.: Buckling and post-buckling analyses for an axially compressed laminated cylindrical shell of FGM with PFRC in thermal environments. Eur. J. Mech. A. Solids **30**, 913–923 (2011)
7. Fallah, A., Aghdam, M.: Nonlinear free vibration and post-buckling analysis of functionally graded beams on nonlinear elastic foundation. Eur. J. Mech. A. Solids **30**, 571–583 (2011)
8. Giunta, G., Crisafulli, D., Belouettar, S., Carrera, E.: Hierarchical theories for the free vibration analysis of functionally graded beams. Compos. Struct. **94**, 68–74 (2011)
9. Hein, H., Feklistova, L.: Free vibrations of non-uniform and axially functionally graded beams using Haar wavelets. Eng. Struct. **33**(12), 3696–3701 (2011)
10. Hsu, J., Lai, H., Chen, C.: Free vibration of non-uniform Euler-Bernoulli beams with general elastically end constraints using Adomian modified decomposition method. J. Sound Vib. **318**, 965–981 (2008)
11. Huang, Y., Li, X.: Bending and vibration of circular cylindrical beams with arbitrary radial nonhomogeneity. Int. J. Mech. Sci. **52**, 595–601 (2010)
12. Huang, Y., Li, X.: A new approach for free vibration of axially functionally graded beams with non-uniform cross-section. J. Sound Vib. **329**, 2291–2303 (2010)
13. Huang, Y., Luo, Q.: A simple method to determine the critical buckling loads for axially inhomogeneous beams with elastic restraint. Comput. Math. Appl. **61**, 2510–2517 (2011)

14. Khalili, S., Jafari, A., Eftekhari, S.: A mixed Ritz-DQ method for forced vibration of functionally graded beams carrying moving loads. Compos. Struct. **92**, 2497–2511 (2010)
15. Kugler, S., Fotiu, P., Murin, J.: The numerical analysis of FGM shells with enhanced finite elements. Eng. Struct. **49**, 920–935 (2013)
16. Li, X., Kang, Y., Wu, J.: Exact frequency equations of free vibration of exponentially functionally graded beams. Appl. Acoust. **74**, 413–420 (2013)
17. Rajasekaran, S.: Differential transformation and differential quadrature methods for centrifugally stiffened axially functionally graded tapered beams. Int. J. Mech. Sci. **74**, 15–31 (2013)
18. Sankar, B.: An elasticity solution for functionally graded beams. Compos. Sci. Technol. **61**, 689–696 (2001)
19. Shahba, A., Rajasekaran, S.: Free vibration and stability of tapered Euler-Bernoulli beams made of axially functionally graded materials. Appl. Math. Model. **36**, 3094–3111 (2012)
20. Şimşek, M., Kocatürk, T., Akbaş, Ş.: Dynamic behavior of an axially functionally graded beam under action of a moving harmonic load. Compos. Struct. **94**, 2358–2364 (2012)
21. Sina, S., Navazi, H., Haddadpour, H.: An analytical method for free vibration analysis of functionally graded beams. Mater. Des. **30**, 741–747 (2009)
22. Singh, K., Li, G.: Buckling of functionally graded and elastically restrained non-uniform columns. Compos. Part B Eng. **40**, 393–403 (2009)
23. Thai, H., Vo, T.: Bending and free vibration of functionally graded beams using various higher-order shear deformation beam theories. Int. J. Mech. Sci. **62**, 57–66 (2012)
24. Wu, L., Wang, Q., Elishakoff, I.: Semi-inverse method for axially functionally graded beams with an anti-symmetric vibration mode. J. Sound Vib. **284**, 1190–1202 (2005)
25. Yaghoobi, H., Torabi, M.: Post-buckling and nonlinear free vibration analysis of geometrically imperfect functionally graded beams resting on nonlinear elastic foundation. Appl. Math. Model. **37**, 8324–8340 (2013)
26. Zhou, D., Cheung, Y.: The free vibration of a type of tapered beams. Comput. Methods Appl. Mech. Eng. **188**, 203–216 (2000)

Chapter 13
Vibrations of Functionally Graded Timoshenko Beams

13.1 Governing Equations

Consider an axially graded Timoshenko beam of length L with a variable cross-section subjected to a constant compressive load P. Following the Timoshenko beam theory and assuming sinusoidal variations of transverse displacement and bending rotation with circular frequency ω, the coupling equations for the transverse displacement $W(\tilde{x})$ and bending rotation $\theta(\tilde{x})$ can be presented as:

$$\frac{d}{dx}\left[E(x)I(x)\frac{d\theta(\tilde{x})}{d\tilde{x}}\right] + kG(\tilde{x})A(\tilde{x})\left[\frac{dW(\tilde{x})}{dx} - \theta(\tilde{x})\right] + \rho(\tilde{x})I(\tilde{x})\omega^2\theta(\tilde{x}) = 0,$$
$$\frac{d}{d\tilde{x}}\left\{kG(\tilde{x})A(\tilde{x})(\frac{dW(\tilde{x})}{d\tilde{x}} - \theta(\tilde{x}))\right\} - \frac{d}{d\tilde{x}}(P\frac{dW(\tilde{x})}{d\tilde{x}}) + \rho(\tilde{x})A(\tilde{x})\omega^2 W(\tilde{x}) = 0. \quad (13.1)$$

In (13.1), $E(\tilde{x})$ and $G(\tilde{x}) = \frac{E(\tilde{x})}{2(1+\nu)}$ denote the modulus of elasticity and rigidity, respectively; $A(\tilde{x})$ is the cross-sectional area, $\rho(\tilde{x})$ is the mass density, $I(\tilde{x})$ is the moment of inertia, ν is the Poisson's ratio and k is the shear correction factor which depends on the shape of the cross-section. The length-wise gradation of material is determined by functions $E(\tilde{x})$ and $\rho(\tilde{x})$, whereas the functions which depend on the cross-section shape are $A(\tilde{x})$ and $I(\tilde{x})$. It is well known that for the determination of the critical buckling load, the quantity ω vanishes, whereas in the case of free vibrations $P = 0$. From (13.1), the equation for determining the critical buckling load can be directly derived

$$\frac{d^2}{d\tilde{x}^2}\left[E(\tilde{x})I(\tilde{x})\frac{d\theta(\tilde{x})}{d\tilde{x}}\right] + P_{cr}\frac{d}{d\tilde{x}}\left[\theta(\tilde{x}) - \frac{1}{kG(\tilde{x})A(\tilde{x})}\frac{d}{d\tilde{x}}[E(\tilde{x})I(\tilde{x})]\frac{d\theta(\tilde{x})}{d\tilde{x}}\right] = 0. \quad (13.2)$$

In the present work, it is assumed that the functions $E(\tilde{x})$, $\rho(\tilde{x})$, $A(\tilde{x})$ and $I(\tilde{x})$ have the following form

Ü. Lepik and H. Hein, *Haar Wavelets*, Mathematical Engineering,
DOI: 10.1007/978-3-319-04295-4_13,

$$\begin{aligned} E(\tilde{x}) &= E_0 E_1(\tilde{x}),\, \rho(\tilde{x}) = \rho_0 \rho_1(\tilde{x}), \\ A(\tilde{x}) &= A_0 A_1(\tilde{x}),\, I(\tilde{x}) = I_0 I_1(\tilde{x}), \end{aligned} \tag{13.3}$$

where $E_0 = E(0)$, $\rho_0 = \rho(0)$, $A_0 = A(0)$ and $I_0 = I(0)$.

Introducing the following dimensionless parameters

$$x = \frac{\tilde{x}}{L},\ \varpi^2 = \frac{\rho_0 A_0 L^4}{E_0 I_0}\omega^2,\ s^2 = \frac{2r^2(1+\nu)}{k},\ r^2 = \frac{I_0}{A_0 L^2} \tag{13.4}$$

and taking into account $P = 0$, the system (13.1) is reduced to:

$$\begin{aligned} &s^2[E_1(x)I_1(x)\theta']' + A_1(x)E_1(x)[W' - \theta] + r^2 s^2 \varpi^2 \rho_1(x) I_1(x)\theta = 0, \\ &\left[E_1(x)A_1(x)(W' - \theta)\right]' + s^2\varpi^2 \rho_1(x)A_1(x)W = 0, \end{aligned} \tag{13.5}$$

where the prime denotes the derivative with respect to x. Eliminating $W(x)$ from the system (13.5), one can obtain

$$F_4(x)\theta^{IV}(x) + F_3(x)\theta'''(x) + F_2(x)\theta''(x) + F_1(x)\theta'(x) + F_0(x)\theta(x) = 0, \tag{13.6}$$

where

$$\begin{aligned} F_0(x) &= \varpi^2\left[r^2(\rho_1 I_1)'' - r^2(\rho_1 I_1)'\frac{(\rho_1 A_1)'}{\rho_1 A_1} + r^2 s^2 \varpi^2 \rho_1^2 \frac{I_1}{E_1} - \rho_1 A_1\right], \\ F_1(x) &= 2r^2\varpi^2(\rho_1 I_1)' + (E_1 I_1)''' - \frac{(\rho_1 A_1)'}{\rho_1 A_1}\left[r^2\varpi^2\rho_1 I_1 + (E_1 I_1)''\right] \\ &\quad + s^2\varpi^2\rho_1\frac{(E_1 I_1)'}{E_1}, \\ F_2(x) &= r^2\varpi^2\rho_1 I_1 + 3(E_1 I_1)'' - 2(E_1 I_1)'\frac{(\rho_1 A_1)'}{\rho_1 A_1} + s^2\varpi^2\rho_1 I_1, \\ F_3(x) &= 3(E_1 I_1)' - \frac{(E_1 I_1)(\rho_1 A_1)'}{\rho_1 A_1}, \\ F_4(x) &= E_1 I_1. \end{aligned} \tag{13.7}$$

The function $W(x)$ can be expressed as:

$$W(x) = F_5(x)\theta'''(x) + F_6(x)\theta''(x) + F_7(x)\theta'(x) + F_8(x)\theta(x), \tag{13.8}$$

where

$$F_5(x)=\frac{E_1I_1}{\varpi^2\rho_1A_1},\ F_6(x)=\frac{2(E_1I_1)'}{\varpi^2\rho_1A_1},$$
$$F_7(x)=\frac{r^2\varpi^2\rho_1I_1+(E_1I_1)''}{\varpi^2\rho_1A_1},\ F_8(x)=\frac{(r^2\rho_1I_1)'}{\rho_1A_1}. \tag{13.9}$$

It is assumed that the fourth derivative of the solution (13.6) is sought in the form:

$$\theta^{IV}(x)=\sum_{i=1}^{2M}a_ih_i(x), \tag{13.10}$$

where a_i are unknown wavelet coefficients. Integrating (13.10) four times and taking into account (2.6) and (2.7), one can obtain

$$\begin{aligned}
\theta'''(x)&=\sum\nolimits_{i=1}^{2M}a_ip_{1,i}(x)+\theta_0''',\\
\theta''(x)&=\sum\nolimits_{i=1}^{2M}a_ip_{2,i}(x)+\theta_0'''x+\theta_0'',\\
\theta'(x)&=\sum\nolimits_{i=1}^{2M}a_ip_{3,i}(x)+\frac{1}{2}\theta_0'''x^2+\theta_0''x+\theta_0',\\
\theta(x)&=\sum\nolimits_{i=1}^{2M}a_ip_{4,i}(x)+\frac{1}{6}\theta_0'''x^3+\frac{1}{2}\theta_0''x^2+\theta_0'x+\theta_0.
\end{aligned} \tag{13.11}$$

In (13.11), the quantities $\theta_0=\theta(0),\ \theta_0'=\theta'(0),\ \theta_0''=\theta''(0),\ \theta_0'''=\theta'''(0)$ can be evaluated from the boundary conditions. Substituting (13.11) into (13.8), the function $W(x)$ obtains the following form:

$$\begin{aligned}
W(x)=&\,F_5(x)\left[\sum\nolimits_{i=1}^{2M}a_ip_{1,i}(x)+\theta_0'''\right]\\
&+F_6(x)\left[\sum\nolimits_{i=1}^{2M}a_ip_{2,i}(x)+\theta_0'''x+\theta_0''\right]\\
&+F_7(x)\left[\sum\nolimits_{i=1}^{2M}a_ip_{3,i}(x)+\frac{1}{2}\theta_0'''x^2+\theta_0''x+\theta_0'\right]\\
&+F_8(x)\left[\sum\nolimits_{i=1}^{2M}a_ip_{4,i}(x)+\frac{1}{6}\theta_0'''x^3+\frac{1}{2}\theta_0''x^2+\theta_0'x+\theta_0\right].
\end{aligned} \tag{13.12}$$

The following boundary conditions are considered:

(i) Clamped—clamped beams (CC)

The boundary conditions for the beam are $W(0)=\theta(0)=0=W(1)=\theta(1)=0$. From (13.11) and (13.12), it follows that $\theta_0=0$ and

$$
\begin{aligned}
&F_5(0)\theta_0''' + F_6(0)\theta_0'' + F_7(0)\theta_0' = 0, \\
&F_5(1)\left[\sum_{i=1}^{2M} a_i p_{1,i}(1) + \theta_0'''\right] + F_6(1)\left[\sum_{i=1}^{2M} a_i p_{2,i}(1) + \theta_0''' + \theta_0''\right] \\
&\quad + F_7(1)\left[\sum_{i=1}^{2M} a_i p_{3,i}(1) + \frac{1}{2}\theta_0''' + \theta_0'' + \theta_0'\right] = 0, \\
&\sum_{i=1}^{2M} a_i p_{4,i}(1) + \frac{1}{6}\theta_0''' + \frac{1}{2}\theta_0'' + \theta_0' = 0.
\end{aligned}
\tag{13.13}
$$

From the system (13.13), the constants θ_0', θ_0'', θ_0''' can be evaluated as follows:

$$
\begin{aligned}
\theta_0' &= \frac{1}{C_0}\sum_{i=1}^{2M} a_i\left[C_1 p_{4,i}(1) + C_2 D_i\right], \\
\theta_0'' &= \frac{1}{C_0}\sum_{i=1}^{2M} a_i\left[C_3 p_{4,i}(1) + C_4 D_i\right], \\
\theta_0''' &= \frac{1}{C_0}\sum_{i=1}^{2M} a_i\left[C_5 p_{4,i}(1) + C_6 D_i\right],
\end{aligned}
\tag{13.14}
$$

where

$$
\begin{aligned}
C_0 &= \frac{1}{2}F_5(0)\left[2F_6(1) + F_7(1)\right] - \frac{1}{3}F_6(0)\left[3F_5(1) + 3F_6(1) + F_7(1)\right] \\
&\quad + \frac{1}{6}F_7(0)\left[3F_5(1) + 2F_6(1) + 2F_7(1)\right], \\
C_1 &= -F_5(0)\left[F_6(1) + F_7(1)\right] + \frac{1}{2}F_6(0)\left[2F_5(1) + 2F_6(1) + F_7(1)\right], \\
C_2 &= \frac{1}{6}\left[F_6(0) - 3F_5(0)\right], \\
C_3 &= F_5(0)F_7(1) - \frac{1}{2}\left[2F_5(1) + 2F_6(1) + F_7(1)\right], \\
C_4 &= \frac{1}{6}\left[6F_5(0) - F_7(0)\right], \\
C_5 &= -F_6(0)F_7(1) + F_7(0)\left[F_6(1) + F_7(1)\right], \\
C_6 &= \frac{1}{2}\left[F_7(0) - 2F_6(0)\right], \\
D_i &= F_5(1)p_{1,i}(1) + F_6(1)p_{2,i}(1) + F_7(1)p_{3,i}(1).
\end{aligned}
\tag{13.15}
$$

(ii) Simply supported beams (SS)

In the case of simply supported beam, the boundary conditions are $W(0) = \theta'(0) = W(1) = \theta'(1) = 0$. In this case, $\theta_0' = 0$ and the quantities θ_0, θ_0'', θ_0''' can be evaluated from the following system:

$$\begin{aligned}
&F_5(0)\theta_0''' + F_6(0)\theta_0'' + F_8(0)\theta_0 = 0,\\
&F_5(1)\left[\sum_{i=1}^{2M} a_i p_{1,i}(1) + \theta_0'''\right] + F_6(1)\left[\sum_{i=1}^{2M} a_i p_{2,i}(1) + \theta_0''' + \theta_0''\right]\\
&\quad + F_8(1)\left[\sum_{i=1}^{2M} a_i p_{4,i}(1) + \frac{1}{6}\theta_0''' + \frac{1}{2}\theta_0'' + \theta_0\right] = 0,\\
&\sum_{i=1}^{2M} a_i p_{3,i}(1) + \frac{1}{2}\theta_0''' + \theta_0'' = 0.
\end{aligned} \tag{13.16}$$

(iii) Cantilever beams (CF)

In this case, one end $x = 0$ is clamped while the other end $x = 1$ is free. The boundary conditions for the beam are $W(0) = \theta(0) = \theta'(1) = 0,\ W'(1) - \theta(1) = 0$. The first two equations will take the following form:

$$\begin{aligned}
&F_5(0)\theta_0''' + F_6(0)\theta_0'' + F_7(0)\theta_0' = 0,\\
&\sum_{i=1}^{2M} a_i p_{4,i}(1) + \frac{1}{6}\theta_0''' + \frac{1}{2}\theta_0'' + \theta_0' = 0.
\end{aligned} \tag{13.17}$$

To get the third equation, one should eliminate $W' - \theta$ from the first equation of the system (13.5):

$$W' - \theta = -\frac{s^2}{A_1 E_1}\left[(E_1 I_1)'\theta' + E_1 I_1 \theta'' - r^2 \varpi^2 \rho_1 I_1 \theta\right]. \tag{13.18}$$

Substituting (13.11) into (13.18) and evaluating the result at $x = 1$ leads to:

$$\begin{aligned}
&(E_1 I_1)'\,|_{x=1}\left[\sum_{i=1}^{2M} a_i p_{3,i}(1) + \frac{1}{2}\theta_0''' + \theta_0''\right]\\
&\quad + E_1(1) I_1(1)\left[\sum_{i=1}^{2M} a_i p_{2,i}(1) + \theta_0''' + \theta_0''\right]\\
&\quad - r^2\varpi^2 \rho_1(1) I_1(1)\left[\sum_{i=1}^{2M} a_i p_{4,i}(1) + \frac{1}{6}\theta_0''' + \frac{1}{2}\theta_0'' + \theta_0'\right] = 0.
\end{aligned} \tag{13.19}$$

(iv) Clamped—simply supported (CS) beams

For the clamped—simply supported beams, the boundary conditions are $W(0) = \theta(0) = W(1) = \theta'(1) = 0$, and the system for finding $\theta_0',\ \theta_0'',\ \theta_0'''$ obtains the following form:

$$\begin{aligned}
&F_5(0)\theta_0''' + F_6(0)\theta_0'' + F_7(0)\theta_0' = 0,\\
&F_5(1)\left[\sum_{i=1}^{2M} a_i p_{1,i}(1) + \theta_0'''\right] + F_6(1)\left[\sum_{i=1}^{2M} a_i p_{2,i}(1) + \theta_0''' + \theta_0''\right]\\
&\quad + F_7(1)\left[\sum_{i=1}^{2M} a_i p_{3,i}(1) + \frac{1}{2}\theta_0''' + \theta_0'' + \theta_0'\right] = 0,\\
&\sum_{i=1}^{2M} a_i p_{3,i}(1) + \frac{1}{2}\theta_0''' + \theta_0'' + \theta_0'.
\end{aligned} \tag{13.20}$$

Substituting (13.10), (13.11) and (13.14) into (13.6), taking into account (12.13) and (13.20), the governing system is derived

$$
\begin{aligned}
a(:)\Big\{ & H(:,l)F_4(l) + \left[P^{(1)}(:,l) + U_1(:)E(l)\right] F_3(l) \\
& + \left[P^{(2)}(:,l) + U_1(:)x_l E(l) + U_2(:)E(l)\right] F_2(l) \\
& + \left[P^{(3)}(:,l) + \frac{1}{2}U_1(:)x_l^2 E(l) + U_2(:)x_l E(l) + U_3(:)E(l)\right] F_1(l) \\
& + \left[P^{(4)}(:,l) + \frac{1}{6}U_1(:)x_l^3 E(l) + \frac{1}{2}U_2(:)x_l^2 E(l) \right. \\
& \left. + U_3(:)x_l E(l)\right] F_0(l)\Big\} = 0, \quad l = 1, \ldots, 2M,
\end{aligned}
\tag{13.21}
$$

where $F_i(l) = F_i(x_l),\ i = 0, \ldots, 4$; $E(l)$ is a unit row vector and

$$
\begin{aligned}
U_1(:) &= \frac{1}{C_0}\left[C_5 P^{(4)}(:,l) + C_6 D(:)\right], \\
U_2(:) &= \frac{1}{C_0}\left[C_3 P^{(4)}(:,l) + C_4 D(:)\right], \\
U_1(:) &= \frac{1}{C_0}\left[C_1 P^{(4)}(:,l) + C_2 D(:)\right], \\
D(:) &= [D_1, D_2, \ldots, D_{2M}]^T.
\end{aligned}
\tag{13.22}
$$

The system (13.21) is linear and homogeneous with regard to a_i and contains the frequency parameter ϖ. For deriving a nontrivial solution, the determinant of system (13.21) must be zero.

13.2 Numerical Examples

Example 13.1: Let us consider buckling of a uniform FG Euler-Bernoulli cantilever beam on elastic foundation. The governing equation for uniform FG Euler-Bernoulli beams resting on the Winkler foundation can be presented as:

$$
\frac{d^2}{dx^2}\left[E(x)I(x)\frac{d^2W}{dx^2}\right] + \lambda\frac{d^2W}{dx^2} + \alpha W = 0, \tag{13.23}
$$

where

$$
x = \frac{\tilde{x}}{L}, \lambda = \frac{PL^2}{E_0 I_0}, \alpha = \frac{kL^4}{E_0 I_0}. \tag{13.24}
$$

Table 13.1 Dimensionless critical buckling loads for the first three modes of uniform cantilever beams with different foundation parameter

α	λ_1	[6]	λ_2	λ_3
0	2.4675	2.4674	22.2095	61.7336
50	8.8617	8.8614	33.0886	64.0980
100	11.9968	11.9964	45.2650	67.7994
200	15.6424	–	59.6947	84.3058

Table 13.2 Dimensionless critical buckling loads for the first three modes of uniform cantilever beams with axial nonhomogeneity

β	λ_1	[6]	λ_2	λ_3
0.5	2.8102	–	27.2420	76.1449
1.0	3.1178	3.1177	31.8907	89.4751
2.0	3.6689	–	40.5098	114.2472
3.0	4.1655	–	48.5651	137.4649

In (13.23) and (13.24) α denotes the normalized restraint stiffness parameter. The critical buckling load is closely related to the end supports of the beam. The expressions for the cantilever beam can be presented as follows [6]:

$$W(0) = 0,\ W'(0) = 0,\ W''(1) = 0,\ W'''(1) + \lambda W'(1) = 0, \qquad (13.25)$$

In Table 13.1, the dimensionless critical buckling loads are calculated for the uniform cantilever beams [2]. With the growth of foundation parameter α, the load also grows significantly. The results calculated for the first mode are compared with ones obtained by Huang and Li: the results are precise.

Next, consider the case of Euler-Bernoulli cantilever beam with axial nonhomogeneity $D = E_0 I_0(1 + \beta x)$. The proposed method calculates the loads for different modes and values of β. The only available result for the present case obtained by FEM in [6] is compared with the one computed by Haar wavelets. The results are similar (Table 13.2).

Example 13.2: Next, consider free vibrations of uniform homogeneous Timoshenko beam. In this case, the Eq. (13.5) take the simpler form:

$$\begin{aligned} &s^2\theta''(x) - (1 - s^2 r^2 \varpi^2)\theta(x) + W'(x) = 0, \\ &W''(x) + s^2\varpi^2 W(x) - \theta'(x) = 0. \end{aligned} \qquad (13.26)$$

The calculations were carried out for different values of shear correction factor k and parameter r^2 if $\nu = 0.3$. The results are presented in Table 13.3 and compared with [3]. The values of frequencies are accurate. In Table 13.3, it is seen that with the increasing value of the shear correction factor or mode, the frequencies also increase.

Table 13.3 Values of the first frequency coefficients in the case of clamped-clamped Timoshenko beam

k	ϖ_1	[3]	ϖ_2	ϖ_3
$r^2 = 0.0025$				
0.50	17.3244	17.3246	39.0385	64.5709
0.85	18.8885	18.8886	44.5263	75.4912
1.00	19.2800	19.2801	46.0286	78.6650
$r^2 = 0.0100$				
0.50	11.7062	11.7067	23.5791	37.4630
0.85	13.9156	13.9160	28.7164	45.9789
1.00	14.5688	14.5691	30.3602	48.7023

Example 13.3: In the case of buckling of non-uniform FG Timoshenko beam the governing equations yield from (13.1), (13.4) and (13.25):

$$\begin{aligned}&\left[E_1(x)I_1(x)\theta'(x)\right]' + \frac{1}{s^2}E_1(x)A_1(x)(W'(x) - \theta(x)) = 0,\\ &\frac{1}{s^2}\left[E_1(x)A_1(x)(W'(x) - \theta(x))\right]' - \lambda W''(x) = 0.\end{aligned} \tag{13.27}$$

Consider the case of the beam with variable cross-section of two types:

$$\begin{aligned}&Case1: A_1(x) = 1 - cx,\ I_1(x) = (1 - cx)^3,\\ &Case2: A_1(x) = (1 - cx)^2,\ I_1(x) = (1 - cx)^4,\end{aligned} \tag{13.28}$$

where c is the taper ratio. It is assumed that the material properties of FG beam vary along the axis of the beam with the following power law relation [17]:

$$E_1(x) = Y_0 x^n + 1,\ Y_0 = E_a/E_z - 1, \tag{13.29}$$

where $E_a = 70\,GPa$ and $E_z = 200\,GPa$ correspond to the Young's moduli of aluminium and zirconia, respectively. The dimensionless critical loads of axially FG tapered clamped-clamped beam with different tapered ratio are presented in Table 13.4 and compared with the ones presented in [17]. It can be observed that the critical load decreases as the taper ratio increases or the moment of inertia decreases.

Example 13.4: Finally, consider functionally graded beams with non-uniform cross-section. Assume that the cross-section, moment of inertia, and the Young's modulus vary according to equations (13.28) and (13.29), respectively. The density of material is assumed to vary according to the following rule:

$$\rho_1(x) = R_0 x^n + 1,\ R_0 = \rho_a/\rho_z - 1, \tag{13.30}$$

Table 13.4 Values of the critical load of axially FG tapered Timoshenko beam ($n = 2$)

c	Case A		Case B	
	λ	[17]	λ	[17]
0.1	10.0278	10.1595	9.4431	9.2468
0.2	8.8903	8.9368	7.1453	7.3377
0.3	7.6742	7.6882	5.4713	5.6402
0.4	6.4024	6.4226	4.0199	4.1590
0.5	5.1330	5.1594	2.7908	2.8976
0.6	3.8925	3.9224	1.7846	1.8590
0.7	2.7070	2.7387	1.0020	1.0466
0.8	1.6114	1.6442	0.4434	0.4650
0.9	0.6612	0.6954	0.1260	0.1187

Table 13.5 The first two natural frequencies of axially FG tapered Timoshenko beam ($n = 2$)

c	Case A			Case B		
	ϖ_1	[17]	ϖ_2	ϖ_1	[17]	ϖ_2
0.1	12.4577	12.4689	26.3780	12.4697	12.4812	26.3922
0.5	11.1699	11.1706	24.6430	11.3140	11.3199	24.8243
0.7	10.1212	10.1036	22.9460	10.5004	10.4579	23.4245

where $\rho_a = 2702$ and $\rho_z = 5700$ denote the density of aluminium and zirconia, respectively.

Table 13.5 shows that with the increase of the tapered ratio, the frequencies drop; however, the changes at the moment of inertia have little influence on the frequencies. Quite the same results were obtained and presented in [17].

13.3 Related Papers

In literature, quite a large number of articles are devoted to the free vibration analysis of homogeneous Euler-Bernoulli and Timoshenko beams; however, considerably less research has been undertaken on the buckling of non-uniform Timoshenko beams and axially functionally graded beams due to the complexity of the problem. For instance, Filipich, Rosales et al used the Fourier series and Timoshenko beam theory to calculate the exact values of natural frequencies [3]. Gunda, Gupta et al. studied a large amplitude free vibration of uniform isotropic Timoshenko beams with geometric nonlinearity and all possible boundary conditions using the finite element method (FEM) and harmonic balance method [4]. An alternative approach was proposed by Liao and Zhong in [10]: nonlinear flexural vibration of tapered Timoshenko beams were calculated with the aid of the differential quadrature method. Mo et al. scrutinized vibration of Timoshenko beams on a nonlinear elastic foundation using a weak form of quadrate element method and noticed that the foundation parameter affects conspicuously the fundamental frequency of the beams [11]; the experiments made by Morfidis proved the case [13]. Yesilce, Demirdag et al. obtained frequencies and mode shapes for free vibration of the multi-span Timoshenko beam with multiple spring-mass systems with a different number of spans and spring-masses in different locations [19]. Huang et al. [7] introduced for investigating the vibration behaviors of axially functionally graded Timoshenko beams an auxiliary function which changes the governing equations with variable coefficients for the deflection and rotation to a single governing equation. Making use of a power series for an unknown function, the single equation turns to a system of linear algebraic equations. Rajasekaran [16] investigated the free vibration analysis of rotating axially FG tapered Timoshenko beams using the differential transformation method and differential quadrature element of the lowest order. The Rayleigh-Ritz method was applied in [14], the finite element method in [5, 12, 17].

Şimşek [18] studied FG Timoshenko beam with pinned-pinned supports with moving harmonic load using Newmark-β method in conjunction with the direct iteration method. Material properties of the beam vary continuously in thickness direction according to a power-law form.

An exact solution for the postbuckling behavior of functionally graded Timoshenko beams was obtained by Rahimi et al. [15]. The material properties are assumed to be graded in the thickness direction according to the power-law distribution. Neglecting the inplane inertia, the three equations of motion are reduced to two nonlinear partial integro-differential equations in terms of the transverse mid-plane deflection and the cross sectional rotation.

Analytical relations between buckling loads of functionally graded Timoshenko (the Youngs modulus and Poissons ratio are assumed to vary through the thickness) and homogeneous Euler-Bernoulli beams were obtained by Li and Batra [8].

Auciello and Ercolano sought a general solution for the analysis of dynamic behaviour of axially loaded non-uniform Timoshenko beams using Rayleigh-Ritz method and assumed that the test functions are orthogonal polynomials [1].

In [9] the Timoshenko beam theory was extended for treating FG beams as well as layered beams. All material properties are arbitrary functions along the beam thickness. A single fourth-order governing partial differential equation is derived and all physical quantities can be expressed in terms of the solution of the resulting equation. The static result of deflection and stress distribution is presented for a cantilever FGB. Also, a dynamic analysis including wave propagation and free vibration is performed.

References

1. Auciello, N., Ercolano, A.: A general solution for dynamic response of axially loaded non-uniform Timoshenko beams. Int. J. Solids Struct. **41**, 4861–4874 (2004)
2. Feklistova, L., Hein, H.: Free vibration and stability analysis of functionally graded Timoshenko beams. In: Pimenta, P. (ed.) 10th World Congress of Computational Mechanics (2012)
3. Filipich, C., Cortinez, M.: Natural frequencies of a Timoshenko beam: exact values by means of a generalized solution. Mecanica Computadonal **14**, 134–143 (1994)

4. Gunda, J., Gupta, R., Janardhan, G., Rao, G.: Large amplitude vibration analysis of composite beams: simple closed-form solutions. Compos. Struct. **93**, 870–879 (2011)
5. Hemmatnezhad, M., Ansari, R., Rahimi, G.: Large-amplitude free vibrations of functionally graded beams by means of a finite element formulation. Appl. Math. Model. **37**, 8495–8504 (2013)
6. Huang, Y., Li, X.: A new approach for free vibration of axially functionally graded beams with non-uniform cross-section. J. Sound Vib. **329**, 2291–2303 (2010)
7. Huang, Y., Yang, L., Luo, Q.: Free vibration of axially functionally graded Timoshenko beams with non-uniform cross-section. Compos. B Eng. **45**, 1493–1498 (2013)
8. Li, S., Batra, R.: Relations between buckling loads of functionally graded Timoshenko and homogeneous EulerBernoulli beams. Compos. Struct. **95**, 5–9 (2013)
9. Li, X.: A unified approach for analyzing static and dynamic behaviors of functionally graded Timoshenko and EulerBernoulli beams. J. Sound Vib. **318**, 1210–1229 (2008)
10. Liao, M., Zhong, H.: Nonlinear vibration analysis of tapered Timoshenko beams. Chaos, Solitons Fractals **36**, 1267–1272 (2008)
11. Mo, Y., Ou, L., Zhong, H.: Vibration analysis of Timoshenko beams on a nonlinear elastic foundation. Tsinghua Sci. Technol. **14**, 322–326 (2009)
12. Mohanty, S., Dash, R., Rout, T.: Parametric instability of a functionally graded Timoshenko beam on Winkler's elastic foundation. Nucl. Eng. Des. **241**, 2698–2715 (2011)
13. Morfidis, K.: Vibration of Timoshenko beams on three-parameter elastic foundation. Comput. Struct. **88**, 294–308 (2010)
14. Pradhan, K., Chakraverty, S.: Free vibration of Euler and Timoshenko functionally graded beams by RayleighRitz method. Compos. B Eng. **51**, 175–184 (2013)
15. Rahimi, G., Gazor, M., Hemmatnezhad, M., Toorani, H.: On the postbuckling and free vibrations of FG Timoshenko beams. Compos. Struct. **95**, 247–253 (2013)
16. Rajasekaran, S.: Free vibration of centrifugally stiffened axially functionally graded tapered Timoshenko beams using differential transformation and quadrature methods. Appl. Math. Model. **37**, 4440–4463 (2013)
17. Shahba, A., Attarnejad, R., Marvi, M., Hajilar, S.: Free vibration and stability analysis of axially functionally graded tapered Timoshenko beams with classical and non-classical boundary conditions. Compos. B **42**, 801–808 (2011)
18. Şimşek, M.: Non-linear vibration analysis of a functionally graded Timoshenko beam under action of a moving harmonic load. Compos. Struct. **92**, 2532–2546 (2010)
19. Yesilce, Y., Demirdag, O.: Effect of axial force on free vibration of Timoshenko multi-span beam carrying multiple spring-mass systems. Int. J. Mech. Sci. **50**, 995–1003 (2008)

Chapter 14
Applying Haar Wavelets in Damage Detection Using Machine Learning Methods

14.1 Dynamic Response of Vibrating Composite Beams with Multiple Delaminations

The basic idea of the present chapter is to establish directly an input-output relationship between the modal responses and the delamination locations/sizes using back-propagation neural networks. In order to employ the approach, it is necessary to obtain the vibration response data of composite beams with different delaminations. A structural dynamic model of a delaminated composite beam is established and the modal response data with various delaminations are obtained through numerical simulations. Let us consider the free vibrations of a composite laminated beam with n non-overlapping delaminations. The geometry of the beam is shown in Fig. 14.1. The delaminated beam is considered as a combination of $3n + 1$ beam sections, connected at the delamination boundaries. Each beam section is treated as a classical Euler-Bernoulli beam model with $L_i >> h_i$ [4, 17, 24]. In the present research, the Euler-Bernoulli beam theory with a constrained mode, rigid connector and bending-extension coupling [4, 24] is considered. The governing equations for the intact beam sections are:

$$D_i \frac{\partial^4 w_i}{\partial x^4} + \rho_i A_i \frac{\partial^2 w_i}{\partial t^2} = 0, \; i = 1, \ldots, 3n + 1, \tag{14.1}$$

where $w_i(x, t)$ is the vertical displacement of the i-th beam section; D_i is the bending stiffness; ρ_i is the density of material; A_i is the cross-sectional area; x is the axial coordinate and t is the time. Using the classical laminate theory [21], the bending stiffness D_i is given by:

$$D_i = D_{11}^{(i)} - \frac{\left(B_{11}^{(i)}\right)^2}{A_{11}^{(i)}}, \tag{14.2}$$

Ü. Lepik and H. Hein, *Haar Wavelets*, Mathematical Engineering,
DOI: 10.1007/978-3-319-04295-4_14,

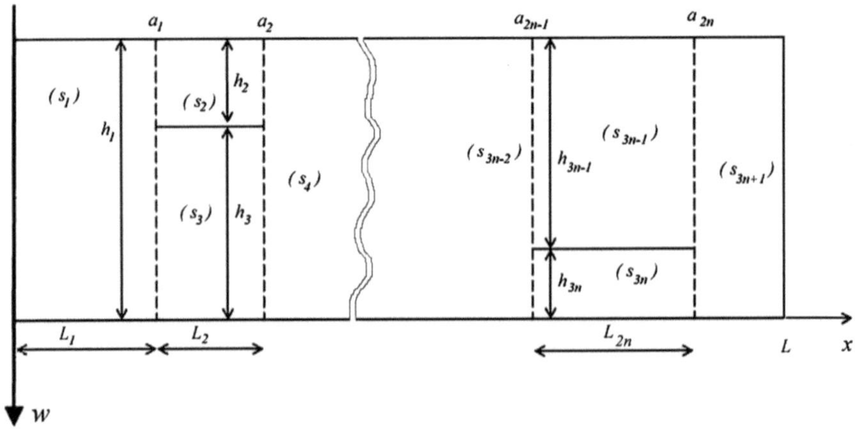

Fig. 14.1 The geometry of the beam

where

$$D_{11}^{(i)} = \frac{b}{3}\sum_{k=1}^{n_i}(\bar{Q}_{11}^{k})_k(z_k^3 - z_{k-1}^3), \tag{14.3}$$

$$B_{11}^{(i)} = \frac{b}{2}\sum_{k=1}^{n_i}(\bar{Q}_{11}^{k})_k(z_k^2 - z_{k-1}^2), \tag{14.4}$$

$$A_{11}^{(i)} = b\sum_{k=1}^{n_i}(\bar{Q}_{11}^{k})_k(z_k - z_{k-1}), \tag{14.5}$$

$$\bar{Q}_{11}^{k} = Q_{11}^{k}\cos^4\varphi + Q_{22}^{k}\sin^4\varphi + 2(Q_{11}^{k} + 2Q_{66}^{k})\sin^2\varphi\cos^2\varphi, \tag{14.6}$$

$$Q_{11} = \frac{E_{11}}{1-\nu_{12}\nu_{21}},\ Q_{22} = \frac{E_{22}}{1-\nu_{12}\nu_{21}},\ Q_{66} = G_{12},\ \nu_{21} = \frac{\nu_{12}E_{22}}{E_{11}}, \tag{14.7}$$

where $D_{11}^{(i)}$ is the bending stiffness, $B_{11}^{(i)}$ is the coupling stiffness, $A_{11}^{(i)}$ is the extensional stiffness of the lamina, b is the width, n_i is the number of plies, ν_{12} and ν_{21} are the longitudinal and transverse Poisson's ratio, respectively, E_{11} and E_{22} are the longitudinal and transverse Young's moduli, respectively, φ is the angle of k-th lamina orientation and z_k and z_{k-1} are the locations of the k-th lamina with respect to the midplane of i-th beam section (Fig. 14.2). According to the constrained model, the beam sections in the delaminated zone are forced to vibrate together, and the governing equations are

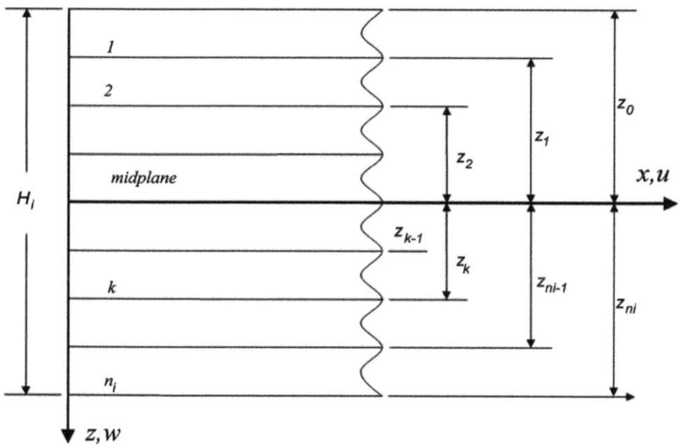

Fig. 14.2 The i-th beam laminate

$$(D_{j1} + D_{j2})\frac{\partial^4 w_j}{\partial x^4} + (\rho_{j1}A_{j1} + \rho_{j2}A_{j2})\frac{\partial^2 w_j}{\partial t^2} = 0,$$
$$j_1 = 2, 5, \ldots, 3n - 1; \; j_2 = 3, 6, \ldots, 3n. \tag{14.8}$$

For free vibrations the solution can be sought in the form

$$w_i(x, t) = W_i(x)\sin(\omega t), \tag{14.9}$$

where ω is the natural frequency and $W_i(x)$ is the mode shape of the i-th beam section. Substituting (14.9) into (14.1), taking into account $x_i = x/L_i$ and eliminating the trivial solution $\sin(\omega t) = 0$, one can obtain the solutions of (14.1) in the following form:

$$W_i(x) = C_{i1}\text{sin}k_i x_i + C_{i2}\text{cos}k_i x_i + C_{i3}\text{sinh}k_i x_i + C_{i4}\text{cosh}k_i x_i, \tag{14.10}$$

where

$$k_i^4 = \frac{\omega^2 \rho_i A_i L_i^4}{D_i} \tag{14.11}$$

and $C_{i1}, \ldots, C_{i4}$ are the arbitrary integrating constants. The solution of (14.8) can be obtained in a similar way. The solution for the beam as a whole is obtained in terms of solutions of all the component beams by enforcing the appropriate boundary and continuity conditions. The Eq. (14.1) is solved in the intact regions and (14.8) in the delaminated regions.

The boundary conditions that can be applied at the supports $x = 0, \; x = L$ are the following. If the beam is clamped at $x = 0$, then $W_1 = 0$ and $W_1' = 0$; if simply supported, then $W_1 = 0$ and $W_1'' = 0$; if free, then $W_1'' = 0$ and $W_1''' = 0$; if guided, then $W_1' = 0$ and $W_1''' = 0$. The analogous boundary conditions can be established at $x = L$. The continuity conditions for deflection, slope and shear force at $x = a_1$ are

$$W_1 = W_2, \; W_1' = W_2', \; D_1 W_1''' = (D_2 + D_3) W_2'''. \tag{14.12}$$

The continuity condition for bending moments at $x = a_1$ can be presented as [4, 19]:

$$\begin{aligned} M_1 &= M_2 + M_3 - \frac{1}{2}P_2(h_1 - h_2) + \frac{1}{2}P_3(h_1 - h_3), \\ M_i &= -D_i W_i'', \; i = 1, 2, 3. \end{aligned} \tag{14.13}$$

The axial forces $P_2, \; P_3$ are established from the compatibility between the stretching/shortening of the delaminated layers and the axial equilibrium [27], which results in

$$\begin{aligned} &\tfrac{P_3 L_2}{EA_1} - \tfrac{P_2 L_2}{EA_2} = (W_1'(a_1) - W_4'(a_2))\tfrac{h_1}{2}, \\ &P_2 + P_3 = 0, \end{aligned} \tag{14.14}$$

where a_1 denotes the coordinate of cross-section between (s_1) and $(s_2)-(s_3)$, whereas a_2 is the coordinate between $(s_2) - (s_3)$ and (s_4) beam sections. Similarly, the continuity conditions are derived at $x = a_2, \ldots, x = a_m; \quad m = 2, \ldots, 2n$. In the case of n delaminations, the boundary conditions and continuity conditions provide $8n + 4$ homogeneous equations for the determination of integration constants.

14.2 Modeling of the Artificial Neural Network (ANN) for Delamination Identification

The back-propagation neural network (BPNN) was used in the present chapter because of its simple approach and precise generalization capability. In order to identify the delamination status using BPNN, a database of sample data consisting of delamination feature indexes and their corresponding delamination status (locations and lengths) should be calculated. The natural frequencies are classically used for damage detection (e.g. [2, 28, 33]). The main reason for the popularity using natural frequencies as damage indicators is that natural frequencies are quite easy to determine with high accuracy [10]. The numerical results of different studies have shown that if nothing is done in order to remove the effects of environment, natural frequencies cannot be used in order to detect damage, whereas mode shapes could be used [5]. Therefore in the present work, the first mode shapes were used to calculate the delamination feature indexes. The main idea is to expand the mode shapes into the Haar wavelet series. Assume that the i-th scaled mode shapes of the delaminated and intact beams are $\mathbf{W}_i^D$ and $\mathbf{W}_i^0$, respectively. The non-dimensional delamination feature index vector of level l (the vector has $2^{l+1} = 2M$ components) is introduced as follows:

$$\mathbf{V}_i^l = (V_{i,1}^l, \ldots, V_{i,2M}^l) = \frac{(\mathbf{W}_{i(2M)}^{*D} - \mathbf{W}_{i(2M)}^{*0})\mathbf{H}_{(2M\times 2M)}^{-1}}{\| (\mathbf{W}_{i(2M)}^{*D} - \mathbf{W}_{i(2M)}^{*0})\mathbf{H}_{(2M\times 2M)}^{-1} \|}, \tag{14.15}$$

where $\| \|$ denotes the Euclidean norm and $\mathbf{W}_{i(2M)}^{*D}$, $\mathbf{W}_{i(2M)}^{*0}$ are the vectors corresponding to the i-th mode shape of the delaminated ($\mathbf{W}_i^D$) and intact ($\mathbf{W}_i^0$) beams, respectively. The components of the vectors are calculated in the collocation points with the aid of cubic splines.

In order to get a stable BPNN, a parametric study on the number of hidden layers and neurons in each hidden layer was performed. The log-sigmoid transfer function was used in this study. The mean-square network error was calculated as follows

$$\mathrm{MSE} = \frac{1}{P}\sum_{p=1}^{P}\sum_{i=1}^{N}(y_i^{(p)} - t_i^{(p)})^2, \tag{14.16}$$

where P is the number of patterns in the training or testing sets, N is the number of output neurons, $y_i^{(p)}$ and $t_i^{(p)}$ denote the actual output and the computed target value in the pattern, respectively. After numerous tests, the network architecture was selected by the following rule: the number of hidden layers is equal to the number of output neurons; the number of neurons in all hidden layers overall is equal to the number of patterns. For every network, the coefficient of multiple determination

$$\mathrm{R}^2 = 1 - \frac{\sum_{p=1}^{P}\sum_{i=1}^{N}(y_i^{(p)} - t_i^{(p)})^2}{\sum_{p=1}^{P}\sum_{i=1}^{N}(y_i^{(p)} - y_m)^2} \tag{14.17}$$

and the variance account for (VAF)

$$\mathrm{VAF} = \left[1 - \frac{var(y-t)}{var(y)}\right]\cdot 100 \tag{14.18}$$

have been calculated [18]. In (14.17) y_m is the mean of output values y and var in (14.18) denotes the variance. The coefficient of multiple determination R^2 shows the closeness of fit, whereas VAF shows the overall performance of the network. Ideally, R^2 is equal to 1 and VAF is equal to 100 %.

14.3 Numerical Examples

Example 14.1: Assume that the symmetric delamination occured in the midplane of the beam. An ANN with 92 neurons placed in a hidden layer was used to study the case. The ANN was trained by six different algorithms (resilient, Fletcher-Revees, Polak-Ribiere, Powell-Beale, Levenberg-Marquardt and Bayesian) using 92 training

Table 14.1 Accuracy of delamination length predictions of beam with simply supported ends

Method	MSE	VAF	R^2	Time(s)
Resilient				
Fr	0.0004	0.9959	0.9948	3.4530
H8	0.0008	0.9910	0.9896	2.2e+1
Fletcher-Reeves				
Fr	0.0010	0.9878	0.9863	1.5470
H8	0.0008	0.9919	0.9893	0.0064
Polak-Ribiere				
Fr	0.0021	0.9754	0.9711	1.7190
H8	0.0011	0.9857	0.9857	9.9530
Powell-Beale				
Fr	0.0015	0.9801	0.9801	1.6250
H8	0.0012	0.9865	0.9838	5.7500
Levenberg-Marquardt				
Fr	0.0002	0.9971	0.9971	6.3750
H8	0.0006	0.9925	0.9921	1.9e3
Bayesian				
Fr	0.0000	0.9995	0.9995	2.2e+2
H8	0.2970	0.9884	0.9857	9.9530

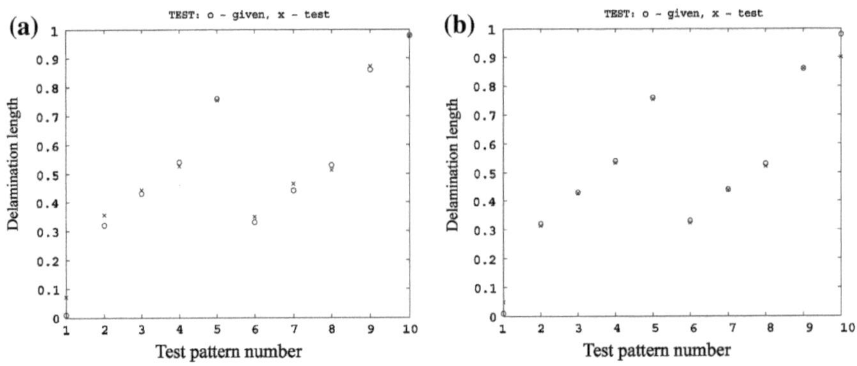

Fig. 14.3 Delamination length prediction using Fletcher-Revees training method, **a**—input: six natural frequencies, **b**—input: eight Haar coefficients

patterns. The accuracy of delamination length prediction is shown in Table 14.1 [12]. The calculations have been repeated with other boundary conditions also. Proceeding from the results, the following conclusions were made. Firstly, the boundary conditions of the beam did not have significant influence on the ANN ability to make predictions of the parameter. Secondly, in most tests the Haar wavelet approach with eight coefficients and ANN trained by Bayesian method made the most accurate predictions for the delaminations smaller than 0.01 units. A visual comparison of the Haar wavelet approach and six natural frequencies is shown in Fig. 14.3. The abscissa exposes the number of the test pattern, the axis of ordinates shows the length of the delamination. The circles indicate the target values, the crosses show the values computed by ANN. Thirdly, in the case of the clamped beam, a mere four Haar coefficients were enough to make accurate predictions. This proves the hypotheses that the signal can be decomposed by Haar wavelets into a certain number of components; the increased number of components results in the decrease of accuracy. Fourthly, the accuracy of predictions depended on the training algorithm. The comparison of the results in the tables shows that different training algorithms work better using either frequencies (resilient, Fletcher-Revees, Levenberg-Marquardt) or Haar coefficients (Polak-Ribiere, Powell-Beale). Finally, in terms of the computing time needed for data generation and ANN training, the overall process was considerably shorter in the case of Haar coefficients. During that time, the ANN passed less than 5000 epochs and reached the performance goal of training equal to 0.0001.

Example 14.2: Let us consider the homogenous beam with two delaminations at an arbitrary height of the beam. The ANN predicted the location of two delaminations occurred within the same layer but at the arbitrary height of the homogenous beam. The network calculated three parameters: the location of the first delamination along the length of the beam L_1/L, the distance between two delaminations L_4/L and the height at which delamination developed h_2/h. The ANN consisted of three hidden layers with 35 neurons on each. The ANN was trained by 140 patterns and tested by eight patterns. The patterns were composed analogically to the previous examples. The tests were conducted for the cantilever (Table 14.2) and a simply supported beam (Table 14.3). In the case of the cantilever beam, all approaches and methods worked effectively. Three desired parameters were predicted with more than 97 % accuracy. However, in the case of resilience, the Polak-Ribiere and Powell-Beale methods, the predictions were provided slightly better by the Haar wavelet approach; conversely, the calculations were a few seconds longer. In the case of the simply supported beam, the boundary conditions of the beam influenced the accuracy of the predictions. The frequency approach was capable of predicting precisely only two parameters, whereas the Haar wavelet approach managed to compute all three parameters with at least 90.00 % accuracy. The later method took significantly less time to accomplish the task. These are the main advantages of the Haar wavelets approach. One of the best training methods for making predictions of three parameters is Bayesian. The accuracy was 100.00 %.

Example 14.3: Composite beam with one delamination in midplane. The $T300/934$ graphite/epoxy beam with a $\left[0^0/90^0\right]_{2s}$ stacking sequence is considered. The dimensions of the 8-ply beam are $127 \times 12.7 \times 1.016\,\mathrm{mm}^3$. The material properties for the lamina are $E_{11} = 134\,\mathrm{GPa}$, $E_{22} = 10.3\,\mathrm{GPa}$, $G_{12} = 5\,\mathrm{GPa}$, $\nu_{12} = 0.33$ and $\rho = 1.48 \times 10^3\,\mathrm{kg/m}^3$. This beam with CF boundary conditions was considered as a benchmark in [4, 6, 21, 23, 25, 34]. The first frequencies calculated for the composite beam with clamped-free boundary conditions (cantilever) are shown in Table 14.4. It was found that a good agreement was obtained between the frequencies calculated by the present model and the experimental [23], analytical [24, 25] and

Table 14.2 Accuracy of predictions of two delaminations placed in the same layer at arbitrary height of the cantilever

Method	MSE	VAF (h_2/h)	VAF (L_1/L)	VAF (L_4/L)	VAF (mean)	Time (s)
Resilient						
Fr	0.0000	0.9999	0.9995	0.9999	0.9994	1.1e+2
H8	0.0000	0.9999	1.0000	0.9996	0.9998	1.2e+2
Fletcher-Reeves						
Fr	0.0001	0.9998	0.9980	0.9853	0.9944	1.5e+2
H8	0.0001	0.9991	0.9973	0.9769	0.9911	1.8e+2
Polak-Ribiere						
Fr	0.0002	0.9985	0.9852	0.9726	0.9854	1.3e+2
H8	0.0000	1.0000	0.9998	0.9998	0.9999	2.0e+2
Powell-Beale						
Fr	0.0000	0.9995	0.9963	0.9951	0.9970	1.2e+2
H8	0.0000	0.9994	0.9999	0.9999	0.9998	1.2e+2
Bayesian						
Fr	0.0000	1.0000	1.0000	1.0000	1.0000	7.2e+3
H8	0.0000	1.0000	1.0000	1.0000	1.0000	7.3e+3

Table 14.3 Accuracy of predictions of two delaminations placed in the same layer at arbitrary height of the beam with simply supported ends

Method	MSE	VAF (h_2/h)	VAF (L_1/L)	VAF (L_4/L)	VAF (mean)	Time (s)
Resilient						
Fr	0.0041	0.9996	0.3177	0.9930	0.7701	1.0e+2
H8	0.0003	0.9571	0.9989	0.9962	0.9841	1.1e+2
Fletcher-Reeves						
Fr	0.0046	0.9911	0.3230	0.9007	0.7396	1.9e+2
H8	0.0005	0.9177	0.9996	0.9980	0.9718	1.6e+2
Polak-Ribiere						
Fr	0.0045	0.9990	0.2820	0.9833	0.7188	1.9e+2
H8	0.0005	0.9196	0.9997	0.9649	0.9533	1.5e+2
Powell-Beale						
Fr	0.0049	0.9998	0.1488	0.9903	0.7130	1.7e+2
H8	0.0012	0.8034	0.9997	0.9981	0.9337	1.8e+2
Bayesian						
Fr	0.0100	0.9965	0.1842	0.2229	0.4678	1.7e+5
H8	0.0000	1.0000	1.0000	1.0000	1.0000	1.2e+4

Table 14.4 Primary frequencies for cantilever composite beam

Delamination length (mm)	Present (Hz)	[24]	[23]	[17]
0.0	82.02	81.88	82.04	81.86
25.4	79.93	80.47	80.13	81.84
50.8	74.36	75.36	75.29	76.81
76.2	65.07	66.14	66.94	67.64
101.6	54.75	55.67	57.24	56.95

Table 14.5 Accuracy of delamination length predictions of the composite beam with clamped ends

Method	MSE	VAF	R^2	Time (s)
Resilient				
Fr	0.0005	NA	NA	1.8750
H8	0.0002	0.6658	0.6125	6.8440
Fletcher-Reeves				
Fr	0.0179	NA	NA	1.5470
H8	0.0001	0.7594	0.7084	4.6720
Polak-Ribiere				
Fr	0.0464	NA	NA	1.5470
H8	0.0003	0.5210	0.4799	8.4370
Powell-Beale				
Fr	0.0500	NA	NA	1.4380
H8	0.0003	0.4553	0.4519	3.7970
Levenberg-Marquardt				
Fr	0.0000	0.9612	0.9568	1.7e+1
H8	0.0001	0.9300	0.8336	1.8430
Bayesian				
Fr	0.0000	0.9991	0.9984	1.5e+3
H8	0.0000	0.9970	0.9965	1.8e+2

FEM results [23]. As could be seen the identification results could be significantly improved by integrating into the process support vector machines [7].

The ANN computed the length of the delamination which occurred in the midplane of the composite beam. The ANN consisted of one hidden layer with 140 neurons in it. The ANN was trained by 140 patterns and tested by ten patterns. The tests were conducted for the cantilever beam and are presented in Table 14.5. The advantage of the Haar wavelet approach is that in the case of a composite beam, it worked with any training method; whereas the frequency approach worked only with the Levenberg-Marquardt and Bayesian training algorithm. Both methods are accurate but time-consuming. The Haar wavelet approach made calculations significantly faster than the frequency approach.

Example 14.4: Let us consider the Euler-Bernoulli beam with one open crack. It is assumed that the crack with depth a locates at distance x_c from the left end of the beam. In the present analysis, the rotational crack compliance is assumed to be dominant and the bending spring constant K_b in the vicinity of crack is given by

$$K_b = \frac{1}{c}, \; c = 5.346 \frac{h}{EI} J\,(\frac{a}{h}), \tag{14.19}$$

where h is the depth of the beam and $J(\frac{a}{h})$ the dimensionless local compliance function. Different forms of the compliance function have been proposed. In the present chapter, the function proposed by Paipetis and Dimarogonas [20] is used

$$\begin{aligned} J\,(\frac{a}{h}) = {} & 1.8624\,(\frac{a}{h})^2 - 3.95\,(\frac{a}{h})^3 + 16.375\,(\frac{a}{h})^4 \\ & - 37.226\,(\frac{a}{h})^5 + 76.81\,(\frac{a}{h})^6 - 126.9\,(\frac{a}{h})^7 \\ & + 172\,(\frac{a}{h})^8 - 143.97\,(\frac{a}{h})^9 + 66.56\,(\frac{a}{h})^{10}. \end{aligned} \tag{14.20}$$

Due to the localized crack effect, the cracked beam can be simulated as two uniform beams joined together by a spring at the crack location. The boundary conditions at the crack location x_c could be expressed as follows

$$\begin{aligned} W_1(x_c+0) &= W_2(x_c-0), \\ \frac{d^2W_1}{dx^2}(x_c+0) &= \frac{d^2W_2}{dx^2}(x_c-0), \\ \frac{d^3W_1}{dx^3}(x_c+0) &= \frac{d^3W_2}{dx^3}(x_c-0), \\ \frac{dW_1}{dx}(x_c+0) + \frac{1}{K_b}\frac{d^2W_1}{dx^2}(x_c-0) &= \frac{dW_2}{dx}(x_c-0). \end{aligned} \tag{14.21}$$

where W_1 and W_2 denote the mode shapes on the left and right beam sections, respectively.

The identification of crack depth and location was performed for the cantilever beam. These quantities have been predicted with the aid of a calculated damage feature index. The first six natural frequencies have been used as an alternative case. The results of two approaches are shown in Table 14.6. As it can be seen in Table 14.6, the method using the Haar wavelets significantly improves the accuracy of predictions.

Table 14.6 Identification of the depth and location of crack in cantilever beam

Exact		Prediction based on six first natural frequencies		Prediction based on Haar Haar transform	
Location	Depth	Location	Depth	Location	Depth
0.18	0.21	0.2148	0.2151	0.1815	0.2109
0.26	0.37	0.3028	0.3776	0.2604	0.3705
0.34	0.37	0.3498	0.3730	0.3496	0.3917
0.50	0.49	0.5010	0.5222	0.5018	0.4923
0.58	0.25	0.6124	0.2659	0.5853	0.2515
0.66	0.25	0.6721	0.2581	0.6603	0.2502
0.74	0.17	0.7674	0.1765	0.7401	0.1729
0.82	0.33	0.8580	0.3397	0.8205	0.3302
0.90	0.25	0.9000	0.2668	0.9001	0.2664

14.4 Related Papers

Vibration-based structural damage detection is a relatively new research topic. According to the structural model, the vibration-based structural damage detection approach can be divided into model-based and signal-based methods [16, 36]. In the model-based methods, the damage locations and severities are established through the comparison of data obtained during the experiments and with the aid of a mathematical model of the structure since structural damages cause changes in the dynamic characteristics. This approach was proposed and experimentally tested by several authors [17, 27]; the review and classification can be found in [4]. The signal-based methods do not use the structural model and detect damage by comparing the structural responses before and after the damage. In [6] the frequency response functions and in [25] the shear horizontal waves, derived from mode conversion of the fundamental Lamb wave, were successfully used to extract the damage detection index.

There exists another classification of the vibration-based structural damage detection approach; it can be divided into traditional and modern-type methods [34]. The traditional methods use only dynamic characteristics of structure, e.g. natural frequencies, mode shapes, modal damping, modal strain energy, etc. The location and severity of damage can be determined by the differences between the structural dynamic characteristics of the damaged and intact structures. The advantage of the method is that insignificant changes in the physical properties of a structure due to damage result in detectable variations in modal parameters (natural frequencies, mode shapes and modal damping) [31]. The main insufficiency of the method is how to extract the most important features from the vibration response with the purpose of damage detection. Furthermore, the method depends on experimental data analysis, and therefore is not convenient for online damage detection. A comprehensive review on vibration-based damage detection methods was presented by Zou et al. [39]. The modern-type damage detection methods are based on real-time measured structural response signals. This approach overcomes the drawbacks of the common non-destructive testing techniques, such as acoustic emission, scanning, X-ray, etc. [36]. The methodology can include neural networks, genetic algorithms, wavelet analysis, etc. Different inverse algorithms for identification of delamination location and length were examined in [37].

The wavelet transform has been applied in structural damage detection by many authors [1, 8, 9, 13, 22, 26, 29, 31, 32]. The advantage of the wavelet-based methods is that they do not require the analysis of the complete structure. The methods are independent of the time-frequency analysis. The wavelet transform decomposes a signal into a set of basis functions. The product of the transform is wavelet coefficients for different scales. Due to time-frequency localization and ability to reveal some hidden parts of data that other signal analysis techniques fail to detect [14, 34]. In wavelet transform analysis, the frequency resolution becomes quite poor in the high frequency region. To overcome this drawback, wavelet packet transform instead of wavelet transform has been applied to the dynamic signals measured from a structure [11, 26, 31, 35].

The back-propagation ANN was trained to predict the delamination size and location from the natural frequencies of the beam in [2, 28]. A combined genetic fuzzy radial basis function neural network for delamination detection using natural frequencies was suggested in [38]. In order to employ the approach, it is necessary to obtain the vibration response data of structures with different damages. For this purpose, a number of authors [2, 30, 33, 35] adopted an appropriate finite element model.

The feasibility of using frequency changes for damage detection is limited since significant damage may cause very small changes in natural frequencies [15, 34]. The fundamental mode shapes are typical features which are frequently used for identification of damage in beams. The sensitivity of mode shapes for damage identification in cantilever beams has been studied and the sensitivity rule has been proposed in [3].

References

1. Bayissa, W., Haritos, N., Thelandersson, S.: Vibration-based structural damage identification using wavelet transform. Mech. Syst. Signal Process. **22**, 1194–1215 (2008)
2. Chakraborty, D.: Artificial neural network based delamination prediction in laminated composites. Mater. Des. **26**, 1–7 (2005)
3. Chukwujekwu, O., Chandrashekhara, K., Jiang, Y.: Delamination prediction in composite beams with built-in piezoelectric devices using modal analysis and neural network. NDT E Int. **30**, 326 (1997)
4. Della, C., Shu, D.: Vibration of delaminated composite laminates: a review. Appl. Mech. Rev. **60**, 1–20 (2007)
5. Deraemaeker, A., Reynders, E., Roeck, G.D., Kullaa, J.: Vibration-based structural health monitoring using output-only measurements under changing environment. Mech. Syst. Sign. Proces. **22**, 34–56 (2008)
6. Fang, X., Luo, H., Tang, J.: Structural damage detection using neural network with learning rate improvement. Comput. Struct. **83**, 2150–2161 (2005)
7. Feklistova, L., Hein, H.: Delamination identification using machine learning methods and Haar wavelets. Comput. Assist. Methods Eng. Sci. **19**, 351–360 (2012)

8. Gentile, A., Messina, A.: On the continuous wavelet transforms applied to discrete vibrational data for detecting open cracks in damaged beams. Int. J. Solids Struct. **40**, 295–315 (2003)
9. Grabowska, J., Palacz, M., Krawczuk, M.: Damage identification by wavelet analysis. Mech. Syst. Signal Process. **22**, 1623–1635 (2008)
10. Hadjileontiadis, L., Douka, E., Trochidis, A.: Fractal dimension analysis for crack identification in beam structures. Mech. Syst. Signal Process. **19**, 659–674 (2005)
11. Han, J., Ren, W., Sun, Z.: Wavelet packet based damage identification of beam structures. Int. J. Solids Struct. **42**, 6610–6627 (2005)
12. Hein, H., Feklistova, L.: Computationally efficient delamination detection in composite beams using Haar wavelets. Mech. Syst. Signal Process. **25**, 2257–2270 (2011)
13. Jin, X., Gupta, S., Mukherjee, K., Ray, A.: Wavelet-based feature extraction using probabilistic finite state automata for pattern classification. Pattern Recogn. **44**, 1343–1356 (2011)
14. Kim, H., Melhem, H.: Damage detection of structures by wavelet analysis. Eng. Struct. **26**, 347–362 (2004)
15. Kim, J., Ryu, Y., Cho, H., Stubbs, N.: Damage identification in beam-type structures: frequency-based method vs mode-shape-based method. Eng. Struct. **25**, 57–67 (2003)
16. Li, Z., Yang, X.: Damage identification for beams using ANN based on statistical property of structural responses. Comput. Struct. **86**, 64–71 (2008)
17. Luo, H., Hanagud, S.: Dynamics of delaminated beams. Int. J. Solids Struct. **37**, 1501–1519 (2000)
18. Martins, J., Tomas, P., Sousa, L.: Neural code metrics: analysis and application to the assessment of neural models. Neurocomputing **72**, 2337–2350 (2009)
19. Mujumdar, P., Suryanarayan, S.: Flexural vibrations of beams with delaminations. J. Sound Vib. **125**, 441–461 (1988)
20. Paipetis, S., Dimarogonas, A.: Anal. Methods Rotor Dyn. Elsevier Applied Science, London (1986)
21. Reddy, J.: Mechanics of laminated composite plates. CRC Press (1997)
22. Rucka, M., Wilde, K.: Application of continuous wavelet transform in vibration based damage detection method for beams and plates. J. Sound Vib. **297**, 536–550 (2006)
23. Shen, M., Grady, J.: Free vibrations of delaminated beams. AIAA J. **30**, 1361–1370 (1992)
24. Shu, D., Della, C.: Free vibration analysis of composite beams with two non-overlapping delaminations. Int. J. Mech. Sci. **46**, 509–526 (2004)
25. Su, Z., Yang, C., Pan, N., Ye, L., Zhou, L.: Assessment of delamination in composite beams using shear horizontal (SH) wave mode. Compos. Sci. Technol. **67**, 244–251 (2007)
26. Sun, Z., Chang, C.: Structural damage assessment based on wavelet packet transform. J. Struct. Eng. **128**, 1354–1361 (2002)
27. Tracy, J., Pardoen, G.: Effect of delamination on the natural frequencies of composite laminates. J. Compos. Mater. **21**, 1200–1215 (1990)
28. Valoor, M., Chandrashekhara, K.: A thick composite-beam model for delamination prediction by the use of neural networks. Compos. Sci. Technol. **60**, 1773–1779 (2000)
29. Wang, Q., Deng, X.: Damage detection with spatial wavelets. Int. J. Solids Struct. **36**, 3443–3468 (1999)
30. Waszczyszyn, Z., Ziemianski, L.: Neural netwoks in mechanics of structures and materials—new results and prospects of applications. Comput. Struct. **79**, 2261–2276 (2001)
31. Wei, Z., Yam, L., Cheng, L.: Detection of internal delamination in multi-layer composites using wavelet packets combined with modal parameter analysis. Compos. Struct. **64**, 377–387 (2004)
32. Yam, L., Yan, Y., Jiang, J.: Vibration-based damage detection for composite structures using wavelet transform and neural network identification. Compos. Struct. **60**, 403–412 (2003)
33. Yam, L., Yan, Y., Jiang, J.: Vibration-based damage detection for composite structures using wavelet transform and neural network identification. Appl. Mech. Rev. **60**, 1–20 (2007)
34. Yan, Y., Cheng, L., Wu, Z., Yam, L.: Development in vibration-based structural damage detection technique. Mech. Syst. Signal Process. **21**, 2198–2211 (2007)

35. Yan, Y., Yam, L.: Detection of delamination damage in composite plates using energy spectrum of structural dynamic responses decomposed by wavelet analysis. Comput. Struct. **82**, 347–358 (2004)
36. Yang, Z., Wang, L., Wang, H., Ding, Y., Dang, X.: Damage detection in composite structures using vibration response under stochastic excitation. J. Sound Vib. **325**, 755–768 (2009)
37. Zhang, Z., Shankar, K., Ray, T., Morozov, E., Tahtali, M.: Vibration-based inverse algorithms for detection of delamination in composites. Compos. Struct. **102**, 226–236 (2013)
38. Zheng, S., Li, Z., Wang, H.: A genetic fuzzy radial basis function neural network for structural health monitoring of composite laminated beams. Expert Syst. Appl. **38**, 11837–11842 (2011)
39. Zou, Y., Tong, L., Steven, G.: Vibration-based model-dependent damage (delamination) identification and health monitoring for composite structures - a review. J. Sound Vib. **230**, 357–378 (2000)

Concluding Remarks

For solving the problems of mathematical calculus by Haar wavelet method at the present time two parallel approaches can be stated; for shortness sake we called them in Sect. 3.7 as the "direct Haar method" and the "operational matrix method".

These methods have much in common. Both of them proceed from the recommendation of Chen and Hsiao that into the Haar series should be expanded the highest derivative appearing in the differential equation to be solved; the lower order derivatives and the function itself will be evaluated by integration. In both methods calculation of the wavelet coefficients is reduced to solution of a system of algebraic equations.

The main difference between these methods is the way of calculating the integrals of wavelets. In the case of the "direct method", these integrations are carried out immediately and the results are presented in the form of universal subprograms. In the case of the "operational matrix method", first a special matrix of integration P is calculated and after that the wanted function is found. Chen and Hsiao presented an ingenious method for evaluating the matrix P (consult Sect. 2.6).

Comparing both approaches we want note that the "direct method"—if we make use of standard programs—is very simple and convenient. It is quite universal, as the same course of solution can be used for a wide range of problems.

As for the "operational matrix method", it seems to be somewhat more complicated and less universal (for solving some problems complementary matrices as the "Haar product matrix" and the "coefficient matrix" must be introduced). The "direct method" is easily applicable for calculating integrals of arbitrary order, but the "operational matrix method" has been used mainly for calculating first order integrals.

In some cases, the "operational matrix method" is more expedient than the "direct method"; such a situation takes place e.g. for solving fractional differential equations (consult Sect. 8.6).

* * *

Ü. Lepik and H. Hein, *Haar Wavelets*, Mathematical Engineering,

DOI: 10.1007/978-3-319-04295-4,

The Haar wavelet method exhibits several advantageous features

(i) Haar wavelets belong to the family of square functions (i.e. they may acquire only the values $0, +1, -1$). Such wavelets are mathematically the most simple when compared with other wavelet families.

(ii) High accuracy is obtained already for a small number of grid points.

(iii) Haar wavelet method is a computer-oriented method; it gives us the possibility to implement standard subprograms.

(IV) The Haar matrices contain many zeros; this makes the Haar transform faster than for other wavelet functions.

(V) Simplicity and small computation costs and a small number of significant wavelet coefficients.

(VI) The method is very convenient for solving boundary value problems, since the boundary conditions are taken into account automatically.

(VII) The Haar method is particularly suitable for systems involving abruptly varying functions.

(VIII) The Haar method is very accurate in detecting singularities of irregular structures and transient phenomena exhibited by the analyzed functions.

(IX) The Haar wavelet technique is highly feasible for damage detection.

(X) Numerical prediction of the shock formation by Haar wavelet method is accompanied only by local oscillations in the vicinity of the shock, whereas e.g. in the Fourier method, the oscillations are presented through the whole domain.

(XI) The Haar method also has some preference before the traditional FEM. In the case of FEM the problem has to be remeshed until sufficient accuracy is reached; this may lead to the low computing efficiency or computing rate. The Haar method utilizes the computation time and storage requirements considerably better.

(XII) In the case of resonance, the Haar method describes accurately the sharp peaks, while for the FEM solution a broad specter of forcing frequencies appear (the signal is smeared out to some region).

By reason of these good features, we can assert that when compared with the classical methods, the Haar wavelet method is much more elegant in theory, more convenient in numerical calculations and, most of all, it is much faster in data processing.

Index

Ü. Lepik and H. Hein, *Haar Wavelets*, Mathematical Engineering,
DOI: 10.1007/978-3-319-04295-4, © Springer International Publishing Switzerland 2014

MIX
Papier aus verantwortungsvollen Quellen
Paper from responsible sources
FSC® C105338

If you have any concerns about our products,
you can contact us on
ProductSafety@springernature.com

In case Publisher is established outside the EU,
the EU authorized representative is:
Springer Nature Customer Service Center GmbH
Europaplatz 3, 69115 Heidelberg, Germany

Printed by Libri Plureos GmbH
in Hamburg, Germany